Revitalizing Health Care Ethics

Stephen Scher · Kasia Kozlowska

Revitalizing Health Care Ethics

The Clinician's Voice

Stephen Scher
McLean Hospital
Department of Psychiatry
Harvard Medical School
Boston, MA, USA

Kasia Kozlowska
The Children's Hospital at Westmead
Disciplines of Child & Adolescent Health
and of Psychiatry
University of Sydney Medical School
Sydney, NSW, Australia

ISBN 978-3-031-78474-3   ISBN 978-3-031-78475-0 (eBook)
https://doi.org/10.1007/978-3-031-78475-0

Cover illustration: Melisa Hasan

This Palgrave Macmillan imprint is published by the registered company Springer Nature Switzerland AG
The registered company address is: Gewerbestrasse 11, 6330 Cham, Switzerland

If disposing of this product, please recycle the paper.

"This is a book that clinicians are sure to enjoy. Scher and Kozlowska integrate findings from ethics, philosophy, psychology, and the social sciences, along with engaging clinical vignettes, to mount a compelling case that 'clinical ethics is inescapably embedded in every clinical interaction and in every clinical process.' In lieu of formal bioethics, the authors stress that clinicians should use their own informal ethical discourse (which they have been using their entire lives) to address problems, ethical or otherwise, in clinical health care: what has gone wrong, why, and what needs to be done to address the needs of our patients? Speaking for myself as a proponent of evidence-based medicine, this approach fully resonates with my experiences teaching resident physicians bedside how to judiciously apply best current research evidence, fully informed by their clinical expertise and their patients' experiences, values, and preferences. The book communicates a deep respect for clinicians and for their existing skills and capacities, including ethical capacities. I hope that the book reaches a wide audience of students, trainees, clinicians, and educators."

—Per Olav Vandvik, MD, PhD, *Chief Executive Officer, MAGIC Evidence Ecosystem Foundation, and Professor, Institute of Health and Society, University of Oslo*

"Scher and Kozlowska have produced a critically important account of everyday ethics and values for health care practitioners. They compellingly demonstrate that clinicians can learn to rely on their own judgment as to what's right, through open and reflective consideration of their personal moral development and ongoing professional commitments. They show that effective treatment, caring, and values are inextricably linked—and without needing to rely on so-called 'experts.' They offer a critical intervention for the reconsideration of traditional reliance on formal bioethical reasoning. *Revitalizing Health Care Ethics: The Clinician's Voice* presents a clear and empathic strategy for clinicians to take responsibility and trust their judgment as they confront ethical questions in the care of patients. This is a must-read for clinicians and bioethicists, students and trainees, their teachers, and anyone concerned about the state of compassionate caregiving and the quality of patient care today."

—Allan M. Brandt, *Harvard University*

"How many of us in clinical health care have sought formal ethical opinions in challenging clinical cases, only to be confused by the length, complexity, and vocabulary of the result? This leaves us with a feeling of inadequacy in regard to our competence in this aspect of our practice. Scher and Kozlowska's book is here to help. It shows how all clinicians develop an ethical dimension that is involved in all their decisions, in most of which the ethical aspects are subliminal. The authors trace how our integration of a sense of right and wrong in childhood serves as the foundation for the sense of ethics that trainee clinicians have as they start their professional journey. Through this journey, dotted with individual, landmark experiences that advance and guide the development of trainees into mature practitioners, clinicians become aware of their own competence to make ethical judgments and how that essential skill is integrated into their everyday work."

—Hugh Martin, MB, BS, FRACS, FRCS (Eng), *pediatric surgeon, Children's Hospital at Westmead; Past President, Australian and New Zealand Association of Paediatric Surgeons; and Past President, Australian and New Zealand Burns Association*

"In their 2018 book *Rethinking Health Care Ethics*, Scher and Kozlowska contrasted the standard, 'top-down' approach to bioethics centered on abstract discussion of principles as applied to problem cases, with the informal, 'bottom-up,' implicitly moral, clinical decision-making practiced every day by interpretive communities of care. Building on this foundation, *Revitalizing Health Care Ethics* uses clear language and well-chosen vignettes to show how individuals' and teams' embedded ethics engage their baseline human capacities and responses, along with their frameworks for taking action. Their account significantly deepens our practical understanding of the ways clinicians come to care well for others and to improve that care. As a valuable and accessible contribution, the book deserves to be widely read."

—John R. Peteet, MD, *Associate Professor of Psychiatry, Harvard Medical School, and former Chair of the Corresponding Committee on Religion, Spirituality, and Psychiatry of the American Psychiatric Association*

"In their new book, Scher and Kozlowska literally *revitalize* health care ethics, seeing clinical ethics as embedded in the lives (*vitae*) and work of health care clinicians. The moral voice of clinicians is restored; proper emphasis is given to informal ethical discourse; and ethics is understood as embedded in everything clinicians say and do. In the 'Scher-Kozlowska revolution,' health care ethics is once again the domain of health care clinicians, where now it should stay."

—Amir Muzur, MD, MA, PhD, *Head, Department of Social Sciences and Medical Humanities, Faculty of Medicine, and Department of Public Health, Faculty of Health Studies, University of Rijeka (Croatia)*

*To our colleagues who have contributed vignettes,
infusing the text with the clinician's voice*

FOREWORD

Revitalizing Health Care Ethics is an important book. It is motivated by the wish to optimize the personal development of health professionals in relation to a culture of care within both health training and service organizations. It is freely available through open access, consistent with its goals of maximizing its reach within health services and helping individuals to achieve their professional goals, maintain their sense of self, and understand and navigate the complexities of the organizations that dominate modern health care today.

Clinicians working at the frontlines of health care bring a wealth of experience and an extensive body of knowledge to engagement in relationships of care. As Scher and Kozlowska point out, the health professional works at the intersection of intellect and affect, and continually encounters the vicissitudes of sickness and death. It can be challenging, to say the least. Much is expected, with the constant challenge of delivering high-quality care, often under time constraints and in a demanding work environment, while maintaining one's moral compass and attending to the deeply human elements of caring for patients and addressing the needs of their families.

As we read this book, we learn how "bioethics" as a separate discipline has arisen relatively recently—only in the last 50 years. This new discipline has promoted the view that there are "right" answers to complex moral problems. Bioethicists may, indeed, debate these matters at great length, though without necessarily yielding a clear or agreed outcome. Such is the nature of intellectual discourse. However, as Scher and Kozlowska point out, this approach doesn't usually work well for the clinician who, confronted with moral problems—whether complex or straightforward—needs to make decisions and decide upon courses of action within limited time frames.

In reading *Revitalizing Health Care Ethics*, I was struck by its being a volume prepared for clinicians, by clinicians. What takes center stage is the experience of health professionals working in real-world conditions where time constraints and resource limitations play a central role in just what decisions get made. It is essential to good practice that health professionals are empowered to act and are supported to learn. In order for these professionals to fulfill the larger community's expectations and demand for quality care, *Revitalizing Health Care Ethics* makes it clear that clinicians need to pay attention to their embodied emotional responses, referred to as *touchstones for learning*. What we often refer to as "gut feelings" are identified as having an important role in understanding patient care and in decision-making, particularly when things don't go "according to plan." These feelings, or touchstones for learning, call for reflection and, when needed, consultation with one's colleagues. Unless this active learning process is encouraged, optimal professional development, as well as patient care, will be compromised.

What the authors also stress is that this professional development occurs against a background of "foundational assets"—including the human capacity for moral understanding and moral action—that students possess at the outset of their professional education and that then continues to develop over the lives of all clinicians. It is the development of these human assets over the life of each clinician that determines the texture of how particular clinicians relate to, and communicate with, patients, families, and colleagues. A rich understanding of everyday ethical discourse is something that health professionals initially *bring* to their education and training in health care and that they then extend and elaborate throughout their professional lives. Scher and Kozlowska mount a powerful argument that effective teaching of clinical ethics needs to respect and build upon this foundation.

In thinking about *Revitalizing Health Care Ethics*, I was reminded of many situations in my own past, including some painfully difficult ones, ranging from my medical student days up through roughly 50 years' experience as a clinician. Like all health professionals, I've been in situations where things didn't go as planned and mistakes were made. On the whole,

I've felt fortunate to have had people around me providing help and support in such situations, which is exactly what is needed in order to learn from, and move past, what has happened. In this regard, I'm sure anyone working in primary care will find that this book brings to life the real world of health care ethics—as the title suggests. We are brought inside the individual struggles of clinicians as they strive to manage the diverse range of ethical challenges always present in health care.

Revitalizing Health Care Ethics helps us see how the potentially daunting and morally challenging domain of clinical care can be nurtured within a framework of supportive, learning networks where each individual matters. Teaching and mentoring can be rewarding tasks to be embraced rather than chores to be avoided. Knowledge isn't an abstraction or simply a set of facts. It is something developed and felt within, where our bodies adjust organically as we go along. Clinicians need to be able to trust the process of moral growth that comes from informal discussions with colleagues and from untold hours working with those seeking care. Like the knowledge, perspectives, and insights gained through the experience of literature, this moral dimension of clinical health care can't be learned from lectures, academic articles, or textbooks.

One of the important dimensions of the book is as a counterbalance to a common, but mistaken, conception of *evidence-based health care*. The originators of this term envisioned that the findings of hard science—the evidence base itself—would be balanced against, and integrated with, the clinical knowledge and experience of health professionals. Too often this second element is left out. As the authors demonstrate, sensitive, humane care—which the larger community expects—is possible only if each clinician never loses sight of his or her *own* body of *clinical* knowledge: the hard-won understanding gained at embodied, emotional, and intellectual levels over each clinician's lifetime of work caring for patients.

In the world of the twenty-first century, we see many jobs being taken over by artificial intelligence, including an increasing range of such interventions in health care. Like many others, however, I often complain about the frustration, whether on the phone or online, of making an inquiry only to find there is nothing but an automated response, a chatbot, providing me with algorithmically generated responses. This book is a good reminder that we shouldn't be overly optimistic about, or threatened by, what such algorithms can potentially achieve in health care. As the authors make abundantly clear, the quality of health care ultimately depends upon the capacity of health professionals to draw on their baseline human experiences and capacities, which come together in each clinician's own professional style and approach to patients, families, and colleagues. It is as discrete, very particular individuals that health professionals connect with, and care for, their patients, who are themselves equally discrete individuals and want to be treated as such. It is their shared experience of the human condition—of human frailties, and of life and death, which will, in the end, claim us all—that brings clinicians and their patients together, creating a sense of connection and solidarity, leavening the experience of illness.

Healing and caring go together. If we want health care systems that are imbued with the human qualities of sensitivity and responsiveness, we would do well to consider the ethical stance of "learning in action" outlined in *Revitalizing Health Care Ethics.*

Anthony Korner, PhD, FRANZCP, MMed (Psychotherapy), MBBS
Associate Professor, University of Sydney Medical School
Director, Westmead Psychotherapy Program for Complex Traumatic Disorders
President, Australian and New Zealand Association of Psychotherapy
Sydney, Australia

PREFACE

In this book, we pursue in greater detail some of the main themes from our earlier, 2018 book, *Rethinking Health Care Ethics*. Of particular note in this context are the notions of interpretive community, informal ethical discourse, and moral engagement. As with that earlier book, we are writing especially to an audience of clinicians rather than bioethicists, though we do believe that this new book has important implications regarding the continuities and discontinuities between the academic ethics of bioethicists and the clinical ethics of health professionals.

The encouragement we received from readers of our 2018 book played an important role in our decision to write another. In this context, we take special note of, and give special thanks to, Ira Bedzow of New York Medical College, Allan Brandt of Harvard University, Robert and Suzanne Fletcher of Harvard Medical School, Helene Helgeland of Oslo University Hospital, Arthur Kleinman of Harvard University, Hugh Martin of the Royal Australasian College of Surgeons, Loyola McLean of the University of Sydney Medical School, Amir Muzur of the University of Rijeka (Croatia), John Peteet of Harvard Medical School, Berge Solberg of the Norwegian University of Science and Technology, Daniel Talmasov of New York University Grossman School of Medicine, Per Olav Vandvik of the University of Oslo, Merrilyn Walton of the University of Sydney School of Public Health, and Marshall Wolf of Harvard Medical School. We especially thank Anthony Korner of the University of Sydney Medical School, who has contributed the Foreword to the book.

In addition to the above, we would also like to thank the many people who have drawn on their own experience, clinical or otherwise, in discussing clinical ethics and in helping us to identify reference materials for this book and potentially relevant clinical phenomena: Wendy Bacon, Megan Chambers, Anamika Chand, Jonathan Chandran, Cher Wei Chuah, Jenny D'Arcy, Earle Durheim, Eileen Gill, Clare Hawkes, Philip Hazell, Karen Hazell-Raine, Anusha Joseph, Yu-Na Kim, Belle Lane, David Lester-Smith, Henry Lu, Vivien Lu, Margaret Mackenzie, Maxine Martin, Katie Mater, Chris Nash, Shane O'Rourke, Roger Parnis, Anne Preisz, Jyoti Raghunandan, Hannah Ryan, Richard Schram, Olivia Schollar-Root, Beverley Turner, Roger West, and Catherine Williamson.

During the writing process, we were blessed with two readers who read and commented on the entire set of chapters and whose astute feedback helped shape this book: the historian of science Allan Brandt and pediatric surgeon Hugh Martin. In addition, other readers have commented on various chapters and sections of the draft manuscript, providing feedback about readability, relevance, and such matters: speech pathologist Janet Baker and social workers Diana Carmody and Sandra Spalding. We thank Anna Vu for her help with the figures.

We thank Robert and Suzanne Fletcher for their permission to reproduce and comment on their emailed response to *Rethinking Health Care Ethics*. We are also particularly grateful to the colleagues who shared their clinical journeys and life experiences with us and the reader by contributing vignettes, all of which remain anonymous. Their stories bring the book's ideas to life and embed the book's analysis in the concrete reality of everyday clinical practice.

Finally, we would like to thank Bruce and Sonja Stephen, who contributed to the cost of this open access publication through donations to the Westmead Medical Research Foundation. We also thank the executive members of the Children's Hospital at Westmead's Psychological Medicine department—Anne Duffy, Katherine Knight, Dina Mahmood, Veena Raghupathy, and Andrea Worth—for supporting publication through the use of funds raised by the Department's teaching activities. Finally, we thank Liam Inscoe-Jones, our editor at Palgrave Macmillan, and Niraimathi Harimoorthi, the editorial assistant who ushered us through the production process.

Sydney, Australia
Sydney, Australia

Stephen Scher
Kasia Kozlowska

CONTENTS

1   Introduction: Looking Back and Looking Forward   1
    The Clinician's Voice   3
    Limitations   9
    References   10

Part I   Foundations

2   Two Modes of Ethics: Formal and Informal   15
    Interpretive Communities   17
    Ways of Life   19
    Moral Modes of Experience   20
    Lived Moral Experience   22
    *Putting It All Together: Implications for Understanding
    Clinical Ethics*   23
    References   25

3   From Outsiders to Insiders   27
    The Dynamics of Community Membership   28
    Becoming a Health Care Professional   31
    *Example Course of Professional Development: Postgraduate
    Training in Internal Medicine*   31
        Physical Examination   34

Therapeutic Knowledge and System Navigation
    for Patient-Centered Care   36
Doing What Comes Naturally   40
References   41

Part II   Forging the Clinician's Voice

4   Building On What's Given   45
In the Beginning   46
A Historical Detour: The Rise of (Bio)ethical Expertise   48
"Mind the Gap"   50
Building On What's Given   51
Roles and Responsibilities   52
References   58

5   Dimensions of Moral Experience   61
The Diversity and Texture of Moral Experience   62
Conceptual Map of Moral Experience   63
    Fairness   63
    Autonomy and Related Concepts   66
    Truthfulness   68
    Trust/Reliance/Respect/Reciprocity   71
    Doing Good/Avoiding Harm   72
    Solidarity   74
Working Moral Frameworks   75
References   76

6   Elements of Action   79
Everyday Conceptual Distinctions Regarding Action   80
The Framework of Prospective Action: Ends, Means,
and the Process of Ongoing Adjustment   82
    Salience and Predictive Coding   83
    The Normal and Morally Expected Course of Events   84
Capacities and Skills That Support the Action Framework   87
    Moral Capacities   87
    Cognitive, Perceptual, and Affective Capacities   87
    Communication Skills   88
    Theory of Mind   88
    Quest for Narrative Coherence   88

Capacity to Recognize Our Limits   88
The Bridge to Action   89
References   89

Part III   Moral Engagement in Clinical Health Care

7   **Touchstones for Learning**   93
A Quick Review: The Foundational Elements of Thought,
Feeling, and Action   93
The Interface of Expectations and Actions: Discrepancies
and the Touchstones for Learning   94
Two Vignettes   99
   La Petite Princesse   99
   The Intractably Puzzled Resident   101
The Dual Role of the Action Framework   103
Nurturing Foundational Assets/the Clinician's Voice   107
References   110

8   **Informal Ethical Discourse and the Touchstone**
   Questions   113
The Touchstone Questions and the Informal Ethical
Discourse of Patient Care   115
   The Touchstone Questions   115
   The Fundamental Role of Moral Judgment   118
Between the Flags   120
Engaging the Whole Person   123
   An Example from Palliative Care   123
   The Socio-Institutional Milieu   125
Clinical Engagement As Moral Engagement   126
References   128

9   **Prospective Action and the Language of the Clinic**   131
Prospective Action   133
   Changing the Course of Care   133
   Program Design   134
   Application of Policy   136
   Conflict Resolution   137
The Compressed Language of Clinical Health Care   140

	Implications	144
	Clinical Problem-Solving	144
	Teaching and Learning Clinical Ethics	145
	Consulting in Clinical Ethics	146
	References	147
10	**Expectations and Discrepancies**	149
	The Advanced Beginner	150
	The Diversity of Skills to Be Learned	152
	Clinical Learning as Inherently Social and Interactive	153
	Clinical Challenges and Cumulative Learning	155
	Variations: Expected and Unexpected	157
	The Pervasive Ethics of Clinical Care	163
	References	164
11	**Two Modes of Clinical Ethics**	167
	A Bifurcated World	168
	Two Modes of Clinical Ethics	172
	Socio-Institutional Ethics	173
	The Socio-Institutional Encounter	174
	Institutional Culture: Systemic Elements	175
	Institutional Culture: Human and Social Elements	177
	Clinical Ethics and the Cycle of Institutional Change	179
	References	180

Part IV   The Clinician's Voice

12	**Nurturing the Clinician's Voice**	185
	The Natural History of the Clinician's Voice	186
	Nurturing (or Not) the Clinician's Voice	187
	Sources of Professional Dysfunction	189
	The Action Cycle and Clinical Training	192
	Baseline Sensitivity to the Human Dimensions	
	of the Clinical Setting	194
	Evidence Is Evidence	196
	The Problem of Time and Workload	197
	References	201

13   Revitalizing Health Care Ethics   203
   Socio-Institutional Ethics   204
   Lifelong Learning   206
   References   207

Index   209

LIST OF FIGURES

Fig. 3.1   Patient Care 2: Physical Examination (from *Internal Medicine Milestones*. © Accreditation Council for Graduate Medical Education 2020)   35

Fig. 3.2   Medical Knowledge 2: Therapeutic Knowledge (from *Internal Medicine Milestones*. © Accreditation Council for Graduate Medical Education 2020)   37

Fig. 3.3   Systems-Based Practice 2: System Navigation for Patient-Centered Care (from *Internal Medicine Milestones*. © Accreditation Council for Graduate Medical Education 2020)   38

Fig. 7.1   The touchstone process: The path from baseline experience to recognition to interpretation/discovery to the touchstone questions. In the real world, the first three steps of the touchstone process—the *touchstones for learning*—are often experienced as one, at least on a conscious level. The three steps of the touchstones for learning are the primary subject matter of this chapter. The *touchstone questions*—the fourth step in this progression—receive only passing attention in this chapter. They are discussed in detail in Chapter 8. © Stephen Scher & Kasia Kozlowska 2025   98

Fig. 8.1   A section of a clinical chart for the vital sign of breathing
            rate. The chart is color-coded into the three zones
            (white, yellow, and red) used in the Between the Flags
            program. As explained in the text, the white zone
            indicates stable observations, the yellow zones indicate
            observations that have shifted outside the homeostatic
            zone, and the red zones indicate observations that reflect
            dangerous levels. © NSW Clinical Excellence Commission
            2020   121
Fig. 11.1   The cycle of institutional change. © Stephen Scher &
            Kasia Kozlowska 2025   178
Fig. 12.1   The action cycle in response to discrepancy: Steps (1)–(4)
            correspond to the *touchstone process*; (1)–(3) correspond
            to the *touchstones for learning*; and (4) corresponds
            to the *touchstone questions*. (5) is the subsequent
            assessment and judgment leading to a specific decision
            to act. (6) is the action itself, (7) the consequences
            of that action, and (8) the observation of those
            consequences, at which point the action in question is
            judged to be (a) complete, (b) unsuccessful or falling
            significantly short of expectations, or (c) ongoing
            but in need of further adjustment. (a) and (b) lead
            to new, prospective actions and new action cycles
            (that begin at Step 5 of the action cycle, after addressing
            the questions regarding prospective action [Text
            Box 9.1]). The ongoing action in (c) continues, subject
            to adjustment, leading to the continuation of another
            cycle for this particular action. © Stephen Scher & Kasia
            Kozlowska 2025   193

LIST OF TEXT BOXES

Text Box 3.1   Internal Medicine Milestones   33
Text Box 6.1   Everyday Conceptual Distinctions Regarding Action   80
Text Box 7.1   Understanding the Discrepant or Unexpected   95
Text Box 7.2   Touchstone Questions   96
Text Box 8.1   Touchstone Questions   116
Text Box 9.1   Questions Regarding Prospective Action   132
Text Box 9.2   The Language of Clinical Health Care   142
Text Box 10.1   Royal Children's Hospital at Melbourne, Nursing
                Guidelines for Chest Drain Management (Main
                Headings and Subheadings Only)   152
Text Box 10.2   Fine-Tuning Basic Skills   160
Text Box 12.1   Sources of Professional Dysfunction   190
Text Box 12.2   A Group Discussion Built Around the Action Cycle
                Depicted in Fig. 12.1   199

LIST OF VIGNETTES

"We need an ethicist"   6
Junior doctor's new mindset during a mental health placement   28
Learning to palpate the gut   34
Learning to navigate the mental health system   36
Setting learning goals through supervision   39
The ethical framework of Horton the elephant   46
Communicating with a hearing-impaired mother   47
Nurse protective of his younger brother   53
Nurse, calm and caring from a young age   55
Medical secretary's visit to a maternity clinic   56
Moving past a changed family structure   63
The gifted first-year flautist auditions for the university orchestra   64
Leaving "too nice" behind   66
Ready to go camping?   67
A young lawyer and the truth   69
A minister's partial truth   71
The new screen door   84
A fractured elbow   86
"La petite princesse"   99
The intractably puzzled resident   101
The gourmand and his duck   105
"I don't have gout"   106
A resident refuses to own a mistake   107
A case-oriented ethics conference   126
A renal fellow's biopsychosocial frame for renal transplants   136

A father's conflict with the clinical team   138
New nurses trying to cope with not yet knowing enough   151
Inserting an in-dwelling Foley catheter   154
Inserting a nasogastric tube   156
Dr. Sassall encounters a mangled leg   158
The "old digger" who stabbed himself with a bread-and-butter knife   158
A discrepant clinical presentation   162
"We need an ethicist"   168
A bioethicist lectures clinicians   168
Physician demoralized by managed care   179
Resident unsupported by her supervisor   187
Oncology fellow supported by his supervisors   187
Crying resident confronts the brutal nature of the health care system   197
Surgeon on learning from each operation   198

CHAPTER 1

Introduction: Looking Back and Looking Forward

Abstract In this chapter we introduce the themes of the book. We highlight the challenges of addressing ethical problems of one kind or another as they arise throughout the clinician's workday. We introduce the reader to some of the distinctions and concepts that we've used in our earlier writings and that are explored in much more detail in this book. The distinction between formal and informal ethical discourse and the notion of interpretive communities are of particular importance here. We contextualize our work in this book as presenting a ground-up approach that builds on health professionals' existing capacities and that fills a gap in the clinical ethics literature.

Keywords Clinical ethics · Clinician's voice · Health care ethics · Informal ethical discourse · Interpretive communities · Lived moral experience · Professional communities

In our 2018 book, *Rethinking Health Care Ethics*, we explored what we perceived as a new approach to clinical ethics, one that built upon health professionals' existing capacities to understand and address ethical problems arising in the course of their work (Scher & Kozlowska, 2018). In this book, we take up where our previous book left off by exploring its central ideas:

© The Author(s) 2025
S. Scher and K. Kozlowska, *Revitalizing Health Care Ethics*,
https://doi.org/10.1007/978-3-031-78475-0_1

- *interpretive communities*: a community of professionals in which trainees become members by mastering the skills, and acquiring the values, that define the profession itself
- *informal ethical discourse*: the language of everyday ethical discourse
- *embedded ethics*: the ethical concepts and values that are interwoven with, and embedded in, all forms of clinical reasoning and clinical interventions
- *touchstones for learning*: the questions that health professionals can ask themselves to identify and explore ethical issues that arise in their work
- *the realities of clinical decision-making*: the inescapable need for health professionals to make ethical decisions under the time constraints of clinical health care (and usually, on the spot)

In particular, we attempt to move our previous analysis ahead by explaining how the phenomena of clinical ethics and the "clinician's voice" build upon and can be traced back through the ongoing life histories—including the professional education and later clinical work—of each trainee and clinician. In looking at clinical health care ethics through the lens of clinicians' lived experience and their personal and professional development over time, rather than through the lens of ethical decision-making as such, we hope to provide a more grounded, clinician-friendly way of understanding the phenomena and demands of clinical ethics. In this context we explore the notion of professional communities (Part I/Foundations: Chapters 2 and 3), look at the ethical and other elements that define the clinician's voice (Part II/Forging the Clinician's Voice: Chapters 4 through 6), and analyze how these elements come into play as trainees and clinicians encounter the ethical challenges, small and large, of clinical health care (Part III/Moral Engagement in Clinical Health Care: Chapters 7 through 11).

It is because this book looks at the *lived moral experience* of trainees and clinicians that we have entitled this new book *Revitalizing Health Care Ethics*, which harkens back to the original sense of *vital*, referring to something "present in living beings or organisms and by which they are animated and their functions maintained" (Oxford English Dictionary Online). As we see it, this lived experience must be recognized, respected, built upon, and strengthened in order to achieve a viable, stable, workable conception of clinical ethics—one that enlivens and empowers clinicians over the course of their careers in health care.

* * *

In discussing our 2018 book, the Swedish bioethicist Pär Segerdahl (2018)—of the Ethics Blog and Uppsala University's Centre for Research Ethics and Bioethics—recognized the foundational character of the clinician's voice. Segerdahl noted how unfortunate it is that

> healthcare ethics is typically not presented as development of abilities we already have as human beings. Instead, it is presented as a new subject. Being ethical is presented as having the specific knowledge of this subject. Ethics then seems to be about reasoning in terms of abstract ethical concepts and principles. It is as if you had come to a course on general moral theory, not healthcare ethics. And since most of us do not know a thing about moral theory, we feel ethically stupid and powerless, and lose our self-confidence.
>
>   However, just as you don't need linguistic theory to speak a language, you don't need moral theory to function ethically. Rather, it is the other way around. It is because we already speak and function ethically that there can be such intellectual activities as grammar and moral theory. Can healthcare ethics be taught without putting the cart before the horse?

Part of our answer to this question was given in *Rethinking Health Care Ethics* and also in a 2020 essay, "Teaching Ethics in Psychiatry: Time to Reset" (Scher & Kozlowska, 2020). A much more probing response is undertaken in this book.

THE CLINICIAN'S VOICE

We believe that the beginning of wisdom is to recognize that clinicians and bioethicists do not think in the same way. To suggest just how and why this is so, we use the email that Robert and Suzanne Fletcher, former Co-Editors of the *Annals of Internal Medicine*, sent to us commenting on *Rethinking Health Care Ethics*. The email highlights the discontinuities between the world of clinical health care ethics and the world of formal bioethics.

> So much of it rings true: the need for clinicians to make yes/no decisions throughout their workdays, whereas others can discuss the pros and cons at their leisure and come to no conclusion; intensely ethical clinical discussions, with large consequences for individual patients, taking place

without relying directly on ethical terminology; and the importance of the 'informal curriculum' (experiences throughout life and exposure to the medical community) in the formation of learners' values.

We recall, with amusement, an ethicist during our sabbatical at [a major U.S. medical school] who said, when we described an everyday clinical decision (one of very many), 'That would take days—even weeks—for me to analyze properly.' He told us if the end results of our decision were harmful, we would have been unethical! Kind of difficult to proceed with our workday under those conditions.

The Fletchers' email succinctly captures what we see as the *clinician's voice*, which is largely absent from what one might call "standard" bioethics. Exploring the email in detail will help to bring out the various dimensions of the clinician's voice and also set the stage for the range of issues to be addressed in the rest of this book. And, as they note, "So much of it rings true."

"the need for clinicians to make yes/no decisions"

A baseline constraint for clinicians in any field of health care is that ethical decisions, large and small, typically need to be made right away, on the run. Also, there's no fudging. In any particular situation, the clinician—whether a doctor or nurse or clinical psychologist or social worker, and whether dealing with a patient or colleague or administrator—needs to make a decision, and take action or not, then and there. Situations, along with occasions for action, evolve; opportunities are either taken or lost.

"throughout their workdays"

Ethical problems of one kind or another, whether involving patients, colleagues, or administrators, arise throughout the workday. Many are minor and relatively narrow in scope, with some scarcely rising to the level of consciousness, if at all. Experienced clinicians, for example, might have encountered dozens of patients who ask for a treatment or other intervention that will provide no benefit, in which case the clinicians may routinely decline the requests with no more than a brief, straightforward explanation. Likewise, clinicians regularly make decisions—sometimes by the minute—about how much information to provide in talking with patients. The level of detail provided in response to any particular patient's "how am I doing?" question may generate wide-ranging answers. The information provided, as well as how it is expressed, will depend upon myriad factors, many of them with an ethical dimension, that differ from patient to patient. And so on and so on, throughout the day.

"whereas others can discuss the pros and cons at their leisure and come to no conclusion"

Philosophers and bioethicists have the time and leisure, as well as the training, to think about ethical problems at length and in great depth. And because of their lack of a direct stake in real-world problem-solving, they can even come to an interim conclusion that more thought or analysis is required. Philosophers and bioethicists also generally frame ethical inquiries differently. For the clinician, the central challenge is to determine the question "what should I do, here, now." By contrast, philosophers and bioethicists are interested more broadly in determining the right answer after all relevant factors are taken into account. What is striking is the discontinuity between the time frame that clinicians have for making most decisions in clinical ethics and the huge allocation of time (many hours, potentially days or weeks) that formal ethical analysis requires in order to reach closure.

"intensely ethical clinical discussions"

What stands out here is the obvious energy that the doctors brought to their engagement with clinical ethics as it arises in daily practice. What this level of engagement makes clear is that when trying to help or advise health professionals on clinical ethics, bioethicists and others should respect and build on this energy. This level of energy and engagement—an indication that the whole person, and not just the intellect, has been brought into play—is ideal for both teaching and learning.

"with large consequences for individual patients"

The Fletchers and their colleagues recognize that their discussions and decisions will often have large consequences for the patients affected. Notably, they have seemingly embraced their capacity to address difficult ethical questions in the care of patients. We are reminded here of a comment by one of our psychiatrist friends, Danielle Vandenberg (1961–2019). After reading our *Rethinking Health Care Ethics*, she mentioned to us that

> I can't tell you how many times my colleagues and I were discussing some problem of ethics and ended the discussion with "We need an ethicist." But now I realize that the ethicist is us!

The points here are twofold. First, health professionals have come to downplay their own capacities to address ethical problems in clinical health care—even difficult ones. And second, as the Fletchers observe, these capacities are usually more than adequate to advance clinicians' understanding of, and to reach closure on, ethical problems as they arise on the ground.

"taking place without relying directly on ethical terminology"

The easiest and most direct way of understanding this comment is that—as we discuss throughout the book—the ethics of everyday health care is interwoven with clinical thinking and clinical interventions, and is expressed in the ordinary clinical language used in each particular setting. The comment brings attention to health care professionals' pervasive use of what we refer to many times throughout this book as *informal ethical discourse*, which is the language of ethics as embedded in everyday discourse and social interactions, professional or otherwise. By contrast, *formal ethical discourse* is the language of ethical theory that is learned in philosophy courses and that dominates bioethics and academic philosophy. It is also the language that bioethicists commonly use in their encounters with health professionals.

As we will argue, clinical ethics is inescapably embedded in every clinical interaction and in every clinical process. And in thinking about these interactions and processes, clinicians use the concrete language of the clinic, always keeping the immediate clinical goals in mind.

"the importance of the 'informal curriculum' (experiences throughout life and exposure to the medical community) in the formation of learners' values"

What we see here is a reminder that professional education in health care does not take place in a vacuum. That is, separate from the formal curriculum, broadly conceived—whether in professional school itself or in subsequent postgraduate training—students/trainees inescapably bring into play their own personal histories, individual values, and capacities to engage in informal ethical discourse. These histories, values, and ethical capacities are shaped or channeled by professional education and training, and refined (and embedded in trainees and, later, clinicians) through clinical experience and the social process of becoming a member of an interpretive (professional) community (Chapter 3). It is a dynamic process in which the formal content of each professional field, the personal characteristics of the student/trainee/clinician, and the repetition and procedural knowledge gained through clinical experience interact to generate an ever-evolving professional self. As we note above, and at various points in the book, this professional self is itself embedded in the broader, overarching framework of an *interpretive community*, which provides content and structure to the professional self, and which also constrains it.

> "We recall, with amusement, an ethicist during our sabbatical at [a major U.S. medical school] who said, when we described an everyday clinical decision (one of very many), 'That would take days—even weeks—for me to analyze properly.'"

The discontinuity between the world of the clinic and the world of the bioethicist is here on vivid display. But the radically different time frames reflect deeper differences. First is the difference in process. As mentioned above, the bioethicist aims to explore the full range of potentially relevant issues, deciding subsidiary issues along the way and bringing him or her, over a period of days or weeks, progressively closer to an ethical conclusion. By contrast, doctors and other health professionals need to act quickly (and often on the spot). What clinicians learn to do is to cut through the potential ethical noise to see what is—for them as clinicians caring for particular patients or dealing with colleagues or administrators in particular circumstances—the path forward. This brings us to the second difference. Whereas the bioethicist wants to obtain the *single correct answer*, clinicians do not have that luxury, and they are also aware that most clinical situations have more than one path forward. Their aim is to provide the *best health care they can within the constraints of their clinical milieus*. By the same token, what they want to achieve is an ethical outcome that is, under the circumstances, the very best they can do— "good enough" but not therefore perfect (Sidebotham, 2017). (For more on this notion of being good enough, see our discussion of the Between the Flags program in Chapter 8.)

Discussion of this portion of the email would be incomplete without mentioning the doctors' reference to their own "amusement." That is, as senior clinicians with decades of clinical experience, they immediately recognized how very odd the bioethicist's stance was. In their particular clinical setting, they almost never had days or weeks.

> "He told us if the end results of our decision were harmful, we would have been unethical!"

Health professionals have no choice but to act as best and as wisely as they can at any particular point in time. It is possible that, in looking back in the way that the bioethicist is suggesting, a decision taken earlier might not have been optimal or may have led to some future harm. But if the clinicians acted as best and as wisely as they could at the time, would the bioethicist have preferred for them to act contrary to what was, at the time, their own best judgment? What clinicians need to do, when looking back at a decision that led to an undesired outcome, is not to see themselves as having acted unethically but to see that their own best judgment led to harm in the future. Why did that happen? What didn't they see or understand? Is there any way to think that they might have seen a way of acting otherwise at the time? *These* are helpful questions that set the stage for ethical growth and improved patient care.

> "Kind of difficult to proceed with our workday under those conditions."

The sense of discouragement in this final sentence is palpable. Health professionals typically work hard, often within strict time constraints and with no actual certainty as to how any particular treatment will work out for any particular patient. If doing one's best under the circumstances is not good enough, what is?

In commenting on the Fletchers' response to *Rethinking Health Care Ethics*, we have attempted to draw out the continuities between that book and this one, with the more specific goal of filling out, in an informal, preliminary way, the range of issues that we address in this book. As before, we are especially trying to help clinicians engage with the ethical dimensions of their work. This book is, in that context, part of an effort to advance clinicians' understanding of, and capacity to use, their own ethical capacities. By the same token, we are hoping to fill an important gap in the literature and practice of bioethics: the importance of the clinician's voice and the ground-up approach. But we are also hoping that bioethicists will take the book's lessons to heart and adjust their own work and goals so that their interventions take into account the clinician's voice and the need to work from the ground up. Bioethicists need to build upon how health professionals *actually* think and act—and not on how they *wish* or *hope* that health professionals would think and act.

LIMITATIONS

Before moving to the next chapter, it is important for us to note what we do *not* expect to accomplish in this book about *clinical* health care ethics. Most importantly, we make no effort to engage *directly* with the various scholarly disputes in the bioethics literature concerning "principlism"— the view elaborated in Beauchamp and Childress's *Principles of Biomedical Ethics* (first published in 1979 and now in its eighth, 2019 edition)—or any of the other overarching approaches to health care ethics associated with virtue ethics, feminist ethics, or any of the value systems embedded in particular religions or cultures. All these approaches (and others) are well established, with extensive literatures of their own. Our interest is in what is missing from these approaches: clinical grounding, with the clinician's voice as its core. We also will not be discussing, except in passing, questions relating to public policy, research ethics, or legal regulation and professional accountability (including licensure, disciplinary frameworks, and negligence/malpractice). All these issues are of major importance but are not the central or immediate focus of this book. Also peripheral in this context are important, pressing questions relating to ecological balance, sustainability, and natural resources (Muzur, 2022; Richie, 2020; Williams et al., 2023). We are concerned, instead, with what has been missing from the bioethics literature and from the literature on professional training in health care. We hope that the book will help health professionals to recognize the legitimacy of their own ethical thinking, to bring that thinking to bear upon their own work, work environment, and culture, to communicate and cooperate more effectively, and to look critically at—and when necessary, act to improve upon—existing practices.

References

Beauchamp, T. L., & Childress, J. F. (2019). *Principles of biomedical ethics* (8th ed.). Oxford University Press.

Beauchamp, T. L., & Childress, J. F. (1979). *Principles of biomedical ethics.* Oxford University Press.

Muzur, A. (2022). The variants of non-mainstream bioethics in Europe: An essay on hopes and disappointments. *Journal of Intercultural Management and Ethics, 5*(1), 5–10. https://doi.org/10.35478/jime.2022.1.02

Oxford University Press (n.d.). Vital (Adjective, sec. I.1.a). In *Oxford English Dictionary Online.*

Richie, C. (2020). Sustainability and bioethics: Where we have been, where we are, where we are going. *New Bioethics, 26*(2), 82–90. https://doi.org/10.1080/20502877.2020.1767920

Scher, S., & Kozlowska, K. (2018). *Rethinking health care ethics.* Palgrave Macmillan.

Scher, S., & Kozlowska, K. (2020). Teaching ethics in psychiatry: Time to reset. *Harvard Review of Psychiatry, 28*(5), 328–333. https://doi.org/10.1097/HRP.0000000000000258

Segerdahl, P. (2018, October 3). *Supporting clinicians to trust themselves.* Ethics Blog, Centre for Research Ethics & Bioethics, Uppsala University. https://ethicsblog.crb.uu.se/2018/10/03/supporting-clinicians-to-trust-themselves/.

Sidebotham, C. (2017). Good enough is good enough. *British Journal of General Practice, 67*(660), Article 311. https://doi.org/10.3399/bjgp17X691409

Williams, P. C. M., Beardsley, J., Isaacs, D., Preisz, A., & Marais, B. J. (2023). The impact of climate change and biodiversity loss on the health of children: An ethical perspective. *Frontiers in Public Health, 10*, Article 1048317. https://doi.org/10.3389/fpubh.2022.104831

PART I

Foundations

Two Modes of Ethics: Formal and Informal

Abstract The shared, lived experience of individuals within a professional community generates the local moral processes that shape and maintain the community, its values, and its commitments. This chapter explores three different conceptions of communities in which shared processes shape the lives and values of community members: *interpretive communities* (at the intersection of literary theory and sociology), *ways of life* (from moral philosophy), and *moral modes of experience* (from anthropology). The chapter also explores the distinction between formal and informal ethical discourse and the two modes of doing ethics—the formal, theoretical (top-down) mode commonly used in academia and the informal (bottom-up) mode embedded in the social and institutional environments in which health professionals work.

Keywords Clinical ethics · Health care ethics · Informal ethical discourse · Interpretive communities · Formal ethical discourse · Local moral processes · Moral modes of experience · Professional communities · Ways of life

In Chapter 1 we introduced the reader to the distinction between formal and informal ethical discourse. This distinction, drawn from moral philosophy (the subfield of philosophy, rather than bioethics per se), captures the difference between the world of philosophical ethics and the world of everyday encounters. The former includes the principles and analytical methods that philosophers and bioethicists bring to bear in addressing moral problems, whereas the latter encompasses the language all of us use on the street, in stores, at work, and at home. Importantly for us, the world of informal ethical discourse encompasses the world of clinical health care and the myriad encounters between health professionals, patients, and families, and between health professionals themselves.

The relationship between formal and informal ethics is that formal ethical discourse is, in effect, an effort to describe, abstract from, and capture informal ethical discourse. In *A Theory of Justice* (1971), the landmark twentieth-century work on moral and political philosophy, John Rawls notes that formal ethical discourse, or "moral philosophy," is best understood as "an attempt to describe our moral capacity" (p. 46). Thus, in relation to the particular focus of his analysis, Rawls notes that a "conception of justice characterizes our moral sensibility when the everyday judgments we make [and the supporting reasons for those judgments] are in accordance with the principles." Put more concretely, the task of *formal* ethical discourse is to understand, and to develop what is, in effect, a descriptive theory of, *informal* ethical discourse.

Formal ethical discourse is *no one's* natural language. It is something that needs to be learned and mastered. As an example, when the first author (SS), in the 1970s, was working toward his PhD in philosophy (specializing in moral and political philosophy), he understood that he was being brought into a discipline—a field of academic study—with its own intellectual standards, methods, and concepts. The questions at the center of the field, including in the subfield of moral philosophy, were ones about justification and knowledge: What is the justification for that? How do we know that? The overriding emphasis was on intellectual clarity and critical reasoning: What does that mean? What is the connection, if one can be shown, between *a* and *b*? As in other fields of academia, there were implicit but baseline expectations and ethical demands *for* intellectual honesty and *against*, for example, plagiarism and fraud. Also, students would develop, over time, a better and better understanding of what distinguished philosophical from other sorts of questions, and of how to answer them; that is, what made certain sorts of answers or theories better or more powerful than others.

Much the same as the above can be said of any academic discipline such as history or political science. But what's easy to miss is that, in the process of becoming a "full member" of any particular academic discipline or professional field, including those of health care, a person *becomes* a philosopher, historian, lawyer, nurse, or doctor. The process of academic or professional training engages the *whole person*—that is, it shapes not just how one thinks but how one feels and acts. Put in another way, it is through the process of academic training or professional education, with each field having its own particular institutions, explicit educational processes, and implicit learning opportunities, that a field's intellectual and behavioral standards, including its ethical standards, come to be embodied, over time, in each trainee.

In order to better understand this process, we will focus on three different conceptions of communities in which shared processes shape the lives and values of community members: *interpretive communities* (at the intersection of literary theory and sociology), *ways of life* (from moral philosophy), and *moral modes of experience* (from anthropology). Along the lines suggested by the renowned sociologist of science Robert K. Merton (Merton, 1963), these three different ways of thinking about communities all emerged toward the end of the twentieth century. All focus on how the shared, lived experience of individuals within a community yields the local moral processes—its informal ethical discourse, including its cognitive, affective, and social dimensions—that shape and maintain the community, its values, and its commitments. What becomes clear is that the language of informal ethical discourse, both in itself and as understood in its broader sociocultural dimensions, is very different from the formal ethical discourse typically used by bioethicists and seen in so much of the bioethics literature.

INTERPRETIVE COMMUNITIES

The notion of *interpretive communities* was first elaborated by the literary theorist Stanley Fish in two books: *Is There a Text in This Class? The Authority of Interpretive Communities* (1980) and *Doing What Comes Naturally* (1989). What makes this idea useful for us is not that it is the one and only way of understanding the institutions of modern medicine but that the notion of an interpretive community provides an especially grounded, tangible way of understanding what it is to be, and to have become, a health professional of any particular sort.

Our point of entry for understanding interpretive communities is the notion of "doing what comes naturally," for it captures—in a few words, a short phrase—the baseline perceptions of health professionals that they *are* doctors, nurses, clinical psychologists, or social workers, and that their work in these capacities flows naturally, as it were, from this sense of themselves, their skill sets, and their responsibilities.

Fish describes an interpretive community "not so much a group of individuals who shared a point of view, but a point of view or way of organizing experience" (1989, p. 141). Each interpretive community's way of organizing experience—its "assumed distinctions, categories of understanding, and stipulations of relevance and irrelevance" (1989, p. 141), along with its shared goals, standards, processes, positions, and social roles—is embedded in its institutional practices and, indeed, in the members of the community themselves. What distinguishes one interpretive community from another—for example, academic fields such as philosophy, history, or chemistry, or professional fields such as medicine, nursing, or law—is each community's distinctive, interconnected sets of practices. Disagreement within interpretive communities is common but is managed (and built upon, generating change and progress) through the shared understandings that constitute each separate community.

Though each member of a particular interpretive community thinks and acts individually, the "interpretive acts" are not themselves individual but communal. They take place within the community's "understood practices and assumptions" (Fish, 1980, p. 306). In this context each member of a particular interpretive community is what Fish calls an "embedded practitioner whose standards of judgment, canons of evidence, or normative measures are extensions" of the community itself (Fish, 1989, p. 144):

> To think within a practice is to have one's very perception and sense of possible and appropriate action issue "naturally"—without further reflection—from one's position as a deeply situated agent. Someone who looks with practice-informed eyes sees a field already organized in terms of perspicuous obligations, self-evidently authorized procedures, and obviously relevant pieces of evidence. (Fish, 1989, pp. 386–387)

In the next chapter (3), we look at how one gains membership in such a community. How does this transformation from generic human being to embedded practitioner, or deeply situated agent, take place?

WAYS OF LIFE

At about the same time that Fish was writing *Is There a Text in This Class?*, the distinguished English philosopher Stuart Hampshire[1] published *Morality and Conflict* (1983) setting forth two separate conceptions of morality—or, in the language used in this book, two separate conceptions of *ethics*. In the first conception, justifying an action or course of conduct involves an appeal to general moral principles—to a rational structure that Hampshire sees as having "the twin principles of utility and justice, however formulated, at its base." This conception embodies what we refer to as formal ethical discourse—the abstract, theoretical, rationalistic approach to ethical evaluation and decision-making seen in so much of contemporary bioethics.

In the second conception, Hampshire sees the justification for an action or course of conduct as "found, not in argument towards a general principle, but in the specification of a complex array of historical realities and causal relations" that have generated, over time, various "conventions, customs and practices" defining a person's "way of life" (p. 5). This conception embodies what we refer to as informal ethical discourse, the knowledge and language of ethics grounded in social experience and concrete interactions with other persons. On this view,

> the complex description of the whole way of life, and of its history, do fill the place occupied in other moral contexts by general principles of utility and justice: that is, the justification [for an action or course of conduct] stops when the interconnections of practices and sentiments within a complete way of life are described. (p. 5)

More broadly, Hampshire sees ways of life as

> coherent totalities of customs, attitudes, beliefs, institutions, which are interconnected and mutually dependent in patterns that are sometimes evident and sometimes subtle and concealed. (p. 6)

As such, "any description of [a way of life] will contain theories used to interpret behaviour and social relations" (p. 5). A way of life will consequently include shared standards for perceiving, understanding, and evaluating both behavior and social conduct. And since these standards are specifically connected with, and integrated into, particular ways of life, Hampshire emphasizes that it is generally a mistake to "abstract [an] activity or practice from its setting in a complete way of life" (p. 6), to "make one-to-one comparisons between activities and practices which are parts of different ways of life" (p. 6), or to make moral assessments based on such comparisons.

In sum, what we see here in Hampshire's notion of ways of life is a specifically ethical analogue of Fish's interpretive communities.

Moral Modes of Experience

In his 1998 Tanner Lecture, delivered at Stanford University, the social anthropologist Arthur Kleinman develops the equivalent of Hampshire's *ways of life* from the ground up, starting at the very baseline of human experience (Kleinman, 1999). He defines that experience

> as the felt flow of *interpersonal* communication and engagements. Those lived engagements take place in a local world. Experience is thoroughly *intersubjective*. It involves practices, negotiations, contestations among others with whom we are connected. It is a medium through which collective and subjective processes interfuse. (pp. 358–359)

We can see here a tangible description of the interactive processes that take place in the "local world" encountered, and lived in, during professional training (and usually thereafter, too) in health care. Notably, it's helpful to keep in mind that these experiences and interactive processes—the practices, negotiations, and contestations with and among others—themselves build upon each person's previous life experiences. The intersubjective experiences of professional training are just one more stage in personal development and, in this case, in becoming a health professional.

For Kleinman, this experience is inherently

> moral because it is the medium of engagement in everyday life in which things are at stake and in which ordinary people are deeply engaged stake-holders who have important things to lose, gain, and to preserve. (Kleinman, 1999, p. 362)

Again, what we see here is a concise, while overarching, description of the interactive environments in which health professionals work, from the beginning of their clinical training and then continuing on for the rest of their professional lives.

As Kleinman notes in *Writing at the Margin: Discourse Between Anthropology and Medicine*, "Each local world is characterized by what matters for its members" (1995, p. 54). And the things at stake on the local level in any field of professional training in health care are various common interests and goals, including the care of patients, the capacity to work within institutional environments such as hospitals and clinics, and mastery of, and advancement within, one's own field. Such matters "thoroughly absorb the attention of participants" (Kleinman, 1999, p. 360) in each field, in each local world, of professional training in health care. It is from this concrete local world that the fundamental professional values and goals of health professionals emerge and within which these values and goals are perceived, articulated, and discussed.

As an anthropologist, Kleinman sees moral issues, in medicine and also other fields, as involving "questions that emerge out of the grounded experiences of sick persons, families, and healers in concrete contexts" (Kleinman, 1995, p. 44). Like Hampshire, Kleinman makes a distinction between the two different conceptions of ethics: the "ethical" (or what we have referred to as *formal ethical discourse*) and the "moral" (or what we have referred to as *informal ethical discourse*). He further notes:

> Whereas ethical discourse is a codified body of abstract knowledge held by experts about "the good" and ways to realize it, moral accounts are the commitments of social participants in a local world about what is at stake in everyday experience. (Kleinman, 1995, p. 45)

Importantly, Kleinman notes that

> the contexts and process of moral life involve more than individuals. They also are based in collective orientations, social resources, and intersubjective action. The moral is actualized not only in subjective space but social transactions over what locally matters. (Kleinman, 1995, p. 45)

In this context the formal ethical discourse of bioethicists involves "abstract articulation and debate over codified values" (Kleinman, 1999, p. 363). It is "reflective and intellectualist, emphasizing cognition . . . and coherence" over affect and behavior (1999, p. 363). From Kleinman's perspective, in "a search for an acontextual objectivity," formal ethical discourse, including that of most bioethicists, provides "a view from nowhere" (1999, p. 363).

By contrast, in the actual lived experience of individuals, their "moral engagements" and "moral sentiments," or what Kleinman refers to as *moral modes of experience*, are interlinked; moral experience is inescapably expressed, in part, via emotional-somatic processes, not just "the mind" (1999, pp. 378–379). Put in another way, in lives lived on the local level, via "local moral processes" (1999, pp. 411, 414)—which is how all of us must live—norms and emotions are linked.

This brings us back to our earlier discussion of Hampshire's *Morality and Conflict*. For Kleinman,

> Each local world is characterized by what matters for its members. . . . In the infrapolitics of family, workplace, and community, . . . the processes of strategic negotiation and interpersonal engagement over what is at stake can be properly regarded as those through which a local moral order (i.e., lived reality) is constituted and expressed. Culture, then, is built up out of the everyday routines and rhythms of social life. It is the medium of social experience. (1995, p. 54)

Kleinman has here described, in elegant terms, what Hampshire described as *ways of life*.

LIVED MORAL EXPERIENCE

Implicit in Fish's notion of *interpretive community* and explicit in Hampshire's *ways of life* and Kleinman's *moral modes of experience* is something that serves as a central theme in this book: the lived moral experience of individuals. For Hampshire, this lived moral experience can be understood as our ongoing immersion within "coherent totalities of customs, attitudes, beliefs, institutions" (Hampshire, 1983, p. 6) that define our *way of life*. For Kleinman, lived moral experience can be understood as our ongoing immersion with *local moral processes* involving the "practices, negotiations, contestations among others with whom we are connected" (Kleinman, 1999, pp. 358–359)—the "everyday routines and rhythms of social life" (Kleinman, 1995, p. 54). For Fish, whose interest is specifically in institutions, or "interpretive communities"—including those in health care—lived moral experience can be understood as involving the shared goals, standards, processes, positions, and social roles that are embedded in a community's institutional practices and in the members of the community themselves (Fish, 1980, 1989).

As we noted in Chapter 1, in *Rethinking Health Care Ethics* we were specifically interested in how health care clinicians *think* about ethics and in the language that clinicians use in that context (Scher & Kozlowska, 2018). But here our interests are much broader. In health care, as elsewhere, the language of informal ethical discourse is embedded within a rich, interactive social environment incorporating (in this case) health professionals' thoughts, feelings, and actions, and reflecting myriad institutional, organizational, and legal factors. All these factors are continually in play, both within the individual and in the interactions between individuals—other health professionals, administrators, patients, and families. By contrast, bioethicists focus almost exclusively on formal ethical discourse and its implications for health professionals and patients. The problem is that such *formal* discourse is disconnected from the concrete, on-the-ground world of clinical health care. And even *informal* ethical discourse—which we consider so important in understanding the thoughts and actions of clinicians on the ground—is only one dimension of the local moral processes that occur in clinical health care.

From an even broader perspective, the here-and-now "moral self" of any particular clinician is the product of the local moral processes of clinical health care, and these local moral processes, or what might call the prevailing "moral ambience" of any particular health care setting, are themselves the product of social, scientific, and historical processes reaching back over time. In this context the formal ethical analyses of bioethics touch only at a distance, and only at a particular instant in time, the local moral processes of clinical health care.

PUTTING IT ALL TOGETHER: IMPLICATIONS FOR UNDERSTANDING CLINICAL ETHICS

What we see in the work of Fish, Hampshire, and Kleinman is a conceptual convergence across the disciplines of literary studies, sociology, philosophy, and anthropology. They present a distinctive way of understanding on-the-ground values and the related local moral processes that are embedded within local institutions and in the intrinsically local actions and interactions of individual actors. Just as importantly, we also see that each local community's moral processes, including its own informal ethical discourse, are embedded in the community members themselves—in their thoughts, feelings, and actions.

With regard to health care, in particular, the values and related moral processes of health care and its various communities (e.g., medicine, nursing, social work, and clinical psychology) are embedded in the individual members of those communities. What is embedded is not just a way of thinking but ways of feeling and acting that, taken together, come to define each health professional's sense of professional identity. Inescapably, each health professional engages their professional work as a *whole person*.

The *informality* of local moral processes should not be taken to imply that such processes are unconscious or unreflective, or even that these processes preclude the use of ethical language or concepts such as rights, duties, or obligations. For example, when a person finds that other participants in a discussion repeatedly cut her off or talk over her, she might well pound on the table and say, "I have a right to talk!!"—a fine use of informal ethical discourse. As another example, parents might often discuss exactly what their obligations are as parents, but again, such discussions are just another dimension of the informal ethical discourse we use every day and all the time.

The formal ethical discourse of philosophers and bioethicists is, by definition, conscious, explicit, and logically structured. But the informal ethical discourse that the rest of us engage in—whether health professionals or simply the man on the street—might be characterized at any particular time as conscious, unconscious, intuitive, reflective, analytical, critical, concrete, mundane, or even puzzling. Nevertheless, the results of these informal moral processes "can be just as probing and revelatory as any formal analysis" (Scher & Kozlowska, 2018, p. 76) (see, e.g., the *Washington Post* essay written by ten-year-old students concerning players' protests [by kneeling] during the playing of the U.S. national anthem before National Football League games [Strauss, 2018]). Much the same can be said of any individual's efforts to clarify or understand moral matters within, and in relation to, the local setting and its particular way of life.

What's at stake in this intersection of literary theory/sociology (Fish, 1980, 1989), philosophical ethics (Hampshire, 1983), and anthropology (Kleinman, 1995, 1999)—not to mention our own previous work in health care ethics and informal ethical discourse (Scher & Kozlowska,

2018)—is of profound importance for health professionals. Most immediately, the implication is that the local moral processes are, at base, sound, legitimate, and worthy of respect, and that these processes are generally adequate for health professionals to understand, probe, analyze, criticize, or creatively respond to whatever occurs in any particular health care setting. That is, in all but especially challenging situations, the "native" moral voice of clinicians—their informal ethical discourse—is more than adequate to determine what's what and to determine what needs to be done.

NOTE

1. Hampshire was Chairman of Princeton University's Department of Philosophy while the first author (SS) was an undergraduate concentrator in philosophy (and took two courses with Hampshire).

REFERENCES

Fish, S. E. (1980). *Is there a text in this class? The authority of interpretive communities.* Harvard University Press.

Fish, S. E. (1989). *Doing what comes naturally: Change, rhetoric, and the practice of theory in literary and legal studies.* Duke University Press.

Hampshire, S. (1983). *Morality and conflict.* B. Blackwell.

Kleinman, A. (1995). *Writing at the margin: Discourse between anthropology and medicine.* University of California Press.

Kleinman, A. (1999). Experience and its moral modes: Culture, human conditions, and disorder [lecture delivered in 1998]. In G. B. Peterson (Ed.), *The Tanner lectures on human values* (Vol. 20, pp. 355–420). University of Utah Press. Cambridge University Press. https://tannerlectures.org/lectures/experience-and-its-moral-modes-culture-human-conditions-and-disorder/

Merton, R. K. (1963). Resistance to the systematic study of multiple discoveries in science. *Archives Européennes de Sociologie. European Journal of Sociology, 4*(2), 237–282. https://doi.org/10.1017/S0003975600000801

Rawls, J. (1971). *A theory of justice.* Belknap Press of Harvard University Press.

Scher, S., & Kozlowska, K. (2018). *Rethinking health care ethics.* Palgrave Macmillan.

Strauss, V. (2018, January 26). Students write about NFL anthem protests: Just because we are fourth graders doesn't mean we don't think about serious things. *Washington Post.* https://www.washingtonpost.com/news/answer-

sheet/wp/2018/01/26/students-write-about-nfl-anthem-protests-just-bec
ause-we-are-fourth-graders-doesnt-mean-we-dont-think-about-serious-thi
ngs/.

CHAPTER 3

From Outsiders to Insiders

Abstract Each field's intellectual and behavioral standards, including its ethical standards, come to be embodied, over time, in each trainee through the process of professional education and clinical training. This chapter examines the social processes by which newcomers, initially outsiders, become part of a professional community. What we see is that the processes of learning and progressive mastery are relational, social, and interactive. Trainees and clinicians, in coming to master this professional milieu, are inescapably also coming to master the moral milieu—the ethical values, standards, and goals—embedded in clinical health care. They become "embedded practitioners whose standards of judgment, canons of evidence, or normative measures are extensions of the community itself."

Keywords Becoming a community member · Clinical ethics · Communities of practice · Embedded practitioner · Health care ethics · Interpretive communities · Legitimate peripheral participation · Local moral processes · Professional communities · Sociocultural practices

As discussed in the last chapter, members of interpretive communities, including the many fields of health care, see themselves as *doing what comes naturally*. But how does that happen? How do *outsiders*, as it were, become *insiders*?

© The Author(s) 2025
S. Scher and K. Kozlowska, *Revitalizing Health Care Ethics*,
https://doi.org/10.1007/978-3-031-78475-0_3

THE DYNAMICS OF COMMUNITY MEMBERSHIP

A key to understanding this process comes from the what the social anthropologist Jean Lave and educational theorist Etienne Wenger refer to *legitimate peripheral participation* (Lave & Wenger, 1991). This expression refers to an immediately recognizable phenomenon that will engage the heart and mind of any health care trainee and any health care professional. So, let's begin.

The opening paragraph of Lave and Wenger's (1991) book, *Situated Learning*, succinctly captures their overall approach. They note that

> learners inevitably participate in communities of practitioners and that the mastery of knowledge and skills requires newcomers to move toward full participation in the sociocultural practices of a community. (p. 29)

In this context the expression *legitimate peripheral participation* refers to the "social process" by which newcomers, initially outsiders, become "part of a community of practice" (p. 29). Learners' peripheral status and their legitimacy are, moreover, interconnected. When potential learners, initially simply outsiders, become legitimate trainees—usually through some sort of entry or admission process—they are enabled to gain access to the community's "sources for understanding" (p. 37). These sources include access to the knowledge and practices of full community members as well as the community's "activities, identities, [and] artifacts" (p. 29)—the "sociocultural practices" referred to above.

The following vignette illustrates how a young doctor adjusted to the challenges he encountered in addressing the mental health issues of patients and also in adjusting to the new mindset required in order to work effectively as a member of an inpatient mental health team.

> As a junior doctor working during the COVID-19 pandemic, I saw an increase in the number of young people presenting with low mood, anxiety, deliberate self-harm, and suicidal thoughts. I did not have a lot of experience dealing with these mental health problems. My first thought was to admit these patients to hospital to keep them safe and to determine how to fix their distress. I was regularly speaking to the mental health team to ask for their help—both in addressing the patients' distress and also, in the process, in addressing my own.
>
> After sitting in on several mental health assessments, however, I learned the importance of getting to know these young people and to understand

their distress and how it was affecting them. I observed how my skilled colleagues in mental health evaluated risk and developed safety plans that resulted in many young people going home with follow-up by community mental health services.

To gain further experience in managing these patients, I decided to undertake a one-year training placement in mental health. Through participation in team interviews—and conducting, in time, many such interviews myself—I gained a better understanding of these patients' current life situations, levels of risk, and ability to engage in safety planning. I also came to appreciate the pros and cons of alternative treatment pathways, including inpatient and outpatient options and the complexities of assessing the safety of managing their mental health in the community.

Even more important, however, was my realization that working with a mental health team required a different mindset from what had been so much valued in pediatric medicine, with its emphasis on fixing every problem that comes along, and as quickly as possible. Instead, I learned (largely through observation) to take a step back, reflect, and assess the overall picture. Since mental health problems don't arise in a vacuum, the way forward, in terms of treatment, is to understand the broader narrative—encompassing the patient's family, school, friends, and socioeconomic milieu. None of this is possible without stepping back and letting the situation develop—sometimes over the course of a single assessment, but more commonly over the days, weeks, or months of working with any particular patient and family.

What we see here is a young doctor's efforts to understand and join into the work of a child and adolescent mental health team. The notion of fixing problems has given way to a broader conception of the problems to be addressed, to a more complex understanding of the patient (and family), and to a more flexible, nuanced approach to identifying the best available treatment pathways and most realistic outcomes.

The case highlights the prominence that Lave and Wenger (1991) give to the *sociocultural* practices of a community. Though this emphasis might come as a surprise, in the process of becoming full community members, learners/trainees/apprentices learn all the following about the community:

> who is involved; what they do; what everyday life is like; how masters talk, walk, work, and generally conduct their lives; how people who are not part of the community of practice interact with it; what other learners are doing; and what learners need to learn to become full practitioners.

It includes an increasing understanding of how, when, and about what old-timers collaborate, collude, and collide, and what they enjoy, dislike, respect, and admire. (Lave & Wenger, 1991, p. 95)

What stands out here is that these various elements, except for "what learners need to learn," all connect up specifically with intersubjective matters—relational, social, and interactive—and not with the substantive knowledge of any particular field. Becoming a full community member is only *in part* an intellectual journey.

The pervasive intersubjectivity of the path toward full community membership brings to mind Kleinman's notion of local moral processes as generated by, and embedded in, the local world defined by any particular community of practice (see Kleinman, 1995, Chapter 2). As would-be, prospective full members of such a community engage both with peers and established members of that community, they inescapably involve themselves in the "processes of strategic negotiation and interpersonal engagement" that characterize any particular "local moral order" (Kleinman, 1995, p. 54)—or, in the present context, community of practice.

It is easy to misunderstand the implications of the above point regarding local moral processes and communities of practice. Bioethicists have long seen the ethical components of health care as separate from the cognitive and skill-based dimensions of health care training and ongoing clinical practice. On that view, it is one thing to be a trained clinician, and it is quite another to understand the ethical dimensions of one's own field of practice. As a measure of the "success" of this approach, many health care professionals have come to see themselves as having little capacity to address ethical problems that arise in patient care or even to contribute to serious ethical discussion. Trainees and clinicians are, on this view, mere consumers of whatever bioethicists, ethics consultants, or clinicians with specific training in ethics can provide. What we will be arguing throughout this book is that this conception of clinical health care ethics, though surely helpful for especially challenging ethical quandaries in contemporary health care, is fundamentally mistaken. The moral processes of clinical health care are interwoven with the social, intellectual, and technical milieu of professional training and professional practice. Trainees and clinicians, in coming to master this milieu, are inescapably also coming to master the moral milieu—the ethical values, standards, and goals—embedded in clinical health care.

BECOMING A HEALTH CARE PROFESSIONAL

For many communities of practice (or interpretive communities, to use Fish's terminology, as described in Chapter 2)—especially those involving any of the professions (Fish, 1980)—the road from neophyte to community member is long and often difficult. At the very outset, the competition for places in professional schools and PhD programs can be intense; admission followed by matriculation is the step that secures the legitimacy of one's peripheral participation in any particular community of (professional) practice. But crossing that threshold is only the first step; years of study and practice lie ahead. Each field has a core of substantive knowledge and of analytical, practical, and social/interpersonal/interactive skills, complemented and supported by professional values and attitudes, that need to be mastered. In the process, the persons themselves—the students/trainees—are transformed.

Within the various fields, or communities of practice, of health care itself, the process of moving toward membership can be tracked in various ways through the published curricula and standards of education/training programs. In the context of this chapter, these curricula and standards can be understood as each field's roadmap for legitimate peripheral participation. In our home state (in Australia) of New South Wales, for example, registered nurses need to have completed a Bachelor of Nursing degree, which includes a progressive "mix of theory, simulation lab work, and clinical placements in various settings," and to then apply for certification by the Nursing and Midwifery Board of Australia (New South Wales Health, 2022). The four-year baccalaureate program, with its mix of activities and its progressively more intensive clinical work, is considered adequate for moving from nursing student to registered nurse (subject to certification). But even then it requires further clinical work and experience for nurses to gain competence beyond the level of "Advanced Beginner." A discussion of nursing figures centrally in Chapter 10, "Moral Engagement in Health Care IV: Expectations and Discrepancies."[1]

EXAMPLE COURSE OF PROFESSIONAL DEVELOPMENT: POSTGRADUATE TRAINING IN INTERNAL MEDICINE

We gain an especially fine-grained understanding of professional development, starting with neophyte and advancing from there, by looking at the practice goals and standards of the United States' Accreditation

Council for Graduate Medical Education (ACGME). In 2000, in an effort to create more determinate goals that would facilitate more useful, reliable assessments of competence and outcomes in medical education and training, the ACGME identified six core competencies for graduate medical education (Edgar et al., 2020):

(1) patient care (including clinical reasoning)
(2) medical knowledge
(3) practice-based learning and improvement (including information management)
(4) interpersonal and communication skills
(5) professionalism
(6) systems-based practice (including health economics and teamwork)

The list is deceptively simple and short. For example, "medical knowledge" encompasses much of the substantive knowledge gained through four years of medical school, which itself is built on the numerous undergraduate science courses required as preparation for medical school. The range of topics is readily apparent via the survey conducted by Clerkship Directors in Internal Medicine—all of whose members have formal roles in teaching internal medicine to medical students (Zakowski et al., 2022). The study's findings are presented in two tables. The first lists 47 clinical conditions and presentations, covering all body systems, that should be included as part of the medical school training in internal medicine. The second lists 17 broad, overarching skills that are unrelated to particular conditions or presentations—for example, diagnostic decision-making, history taking, interpretation of various sorts of tests and investigations, and health equity—that medical students are expected to master before they even begin their post–medical school training (Zakowski et al., 2022).

This background of knowledge and skills is then extended through residency programs in internal medicine. The ACGME's Milestones project (2020) describes exactly what is required of residents in internal medicine as they progress through their three-year programs.[2] The milestones themselves (see list in Text Box 3.1) are organized under the six core competencies mentioned above. A careful look at the milestones in internal medicine will provide a useful picture of the progression of trainees' legitimate peripheral participation within medicine—leading, in the end, to full community membership. For the purpose of illustrating this progression, we will look at three of those milestones:

- Patient Care 2: Physical Examination—as an example of a process that needs to be progressively mastered
- Medical Knowledge 2: Therapeutic Knowledge—as an example of the substantive medical knowledge that needs to be progressively mastered
- Systems-Based Practice 2: System Navigation for Patient-Centered Care—as an example of social and communication processes that need to be progressively mastered

---

Text Box 3.1 Internal Medicine Milestones

Patient Care

1. History
2. Physical Examination
3. Clinical Reasoning
4. Patient Management—Inpatient
5. Patient Management—Outpatient
6. Digital Health

Medical Knowledge

1. Applied Foundational Sciences
2. Therapeutic Knowledge
3. Knowledge of Diagnostic Testing

Systems-Based Practice

1. Patient Safety and Quality Improvement
2. System Navigation for Patient-Centered Care
3. Physician Role in Health Care Systems

Practice-Based Learning and Improvement

1. Evidence-Based and Informed Practice
2. Reflective Practice and Commitment to Personal Growth

Professionalism

1. Professional Behavior
2. Ethical Principles
3. Accountability/Conscientiousness
4. Knowledge of Systemic and Individual Factors of [i.e., that affect] Well-Being

Interpersonal and Communication Skills

1. Patient- and Family-Centered Communication
2. Interprofessional and Team Communication
3. Communication within Health Care Systems

Physical Examination

The milestones for Physical Examination are a good place to start, for the progressive changes from one level of expected competence to another (bolded in Fig. 3.1) are so readily apparent. Note that each stage reflects the wealth of experience and learning that occurred at each previous stage. One stage of mastery leads to another set of challenges, and as the new challenges are mastered, they become part of each resident's professional repertoire—something that the resident can do competently, easily, and, while attending carefully, without thinking.

The very beginning—and very foundation—of this ongoing process of growth and change can be seen in the simple example of a medical student who is learning to palpate the gut.

When I was in medical school, my classmates and I were introduced to palpation in small groups, with senior clinicians explaining and then demonstrating the technique. During the demonstration that I was part of, the clinician mentioned that while she was palpating the gut, she was also thinking what might explain the patient's symptoms, what aspects of the patient's history might be relevant, and so on. The universal response by us students was to be aghast that the clinician could be thinking about such things while exploring the patient's gut through what was, for us, a nearly opaque process. How do you obtain information from that?

Before too long, of course, palpating the gut would come to be a technique that the medical students would use effortlessly—naturally—in their physical examination of patients. And so it is, time after time, year after year, as medical students become interns and residents, and then (at last) full members of the interpretive community of practicing physicians—medicine's community of practice.

We see this progression at work in the internal medicine milestone for Physical Examination. The first stage is simply to master the baseline skills involved in physical examination. At the start, feeling a liver edge is one thing; knowing exactly what you're looking for and why is quite another. As the progression of skills in physical examination indicates, having a hypothesis gives shape to the examination. Exploring (and sorting out) hypotheses becomes even more challenging in the case of complex patient presentations. Still more challenging are cases that require advanced techniques or approaches to generate subtle findings that can inform diagnosis and treatment. Finally, as residents advance through these stages, they

Patient Care 2: Physical Examination				
Level 1	Level 2	Level 3	Level 4	Level 5
Performs a general physical examination while attending to patient comfort and safety	Performs a **hypothesis-driven** physical examination for a **common patient presentation**	Performs a hypothesis-driven physical examination for a **complex patient presentation**	**Uses advanced maneuvers to elicit subtle findings**	**Models** effective evidence-based physical examination technique
Identifies common abnormal findings	**Interprets** common abnormal findings	**Identifies and interprets uncommon** and **complex** abnormal findings	**Integrates subtle physical examination findings to guide diagnosis and management**	**Teaches the predictive values of the examination findings** to guide diagnosis and management

Comments:

Not Yet Assessable
Not Yet Completed Level 1

Fig. 3.1 Patient Care 2: Physical Examination (from *Internal Medicine Milestones.* © Accreditation Council for Graduate Medical Education 2020)

will likely also become better able to model and teach physical examination skills to less advanced residents. Throughout this progression, each stage—once mastered—serves as the foundation for the next. By the same token, with repetition the skills at each stage become progressively more easy to perform. In time, they become an effortless part of each resident's foundational repertoire of skills.

The full panoply of physical examination skills (and of bedside medicine, more generally) is helpfully summarized in the Stanford Medicine 25 initiative (Stanford Medicine, 2024).

Therapeutic Knowledge and System Navigation for Patient-Centered Care

After the above, somewhat detailed review of the ACGME milestone for Physical Examination, a brief look at the milestones for Therapeutic Knowledge (under Medical Knowledge) and for System Navigation for Patient-Centered Care (under Systems-Based Practice) (see Text Box 3.1) should suffice for our purposes here.

For each milestone, and akin to the milestone for Physical Examination, simpler, easier tasks serve as the foundation for a series of progressively more complex, challenging tasks. But, whereas the Physical Examination milestone focused on a particular *process*—physical examination—the Therapeutic Knowledge milestone focuses on medical *knowledge* per se (see Fig. 3.2), and the System Navigation for Patient-Centered Care milestone focuses on the institutional context of providing medical care and on the *social and communication skills* required to ensure that patients receive the medical care that they need (see Fig. 3.3). These system-navigation skills are especially important in view of the ever-increasing complexities of today's health care system (see Fig. 3.3).

The vignette below highlights the complexity of this task.

When I joined the mental health team in the Emergency Department as a nurse, I found that the task of learning how to navigate the broader mental health system was unexpectedly challenging. As a reference for clinicians (especially for young clinicians like me), the services available in the community were documented on a sheet hung on the wall, but the process of determining the appropriate service and making a successful referral was deceptively difficult. Following the risk assessment itself—most children presented with suicidal ideation, self-harm, or behavioral escalation in the context of family conflict—negotiation of what was acceptable to the child and family was critical. Some families insisted on admission (whether or not admission was indicated or even possible); others accepted referrals to community services; and others refused any professional engagement. And then came the negotiation with the community services themselves. Would the service accept this child, or would they push back and argue that the child did not meet their criteria? What information did I need to present to any particular service, and what details would be considered a needless distraction, brought into play only because of my own inexperience? Treatment paths opened or closed depending on the success of this negotiation. Fortunately, senior team members were always available for support and brainstorming, especially when the best path forward was uncertain. The experience they brought to these encounters also helped to give life and substance to the inert list of community services hanging on the wall.

The differences between the milestones for Therapeutic Knowledge and for System Navigation for Patient-Centered Care mask a pervasive commonality. For each milestone, the process of learning and progressive mastery is relational, social, and interactive. When medical students and

Medical Knowledge 2: Therapeutic Knowledge				
Level 1	Level 2	Level 3	Level 4	Level 5
Explains the scientific basis for common therapies	Explains the indications, contraindications, risks, and benefits of common therapies	Integrates knowledge of therapeutic options in patients with comorbid conditions, multisystem disease, or uncertain diagnosis	Integrates knowledge of therapeutic options within the clinical and psychosocial context of the patient to formulate treatment options	Demonstrates a nuanced understanding of emerging, atypical, or complex therapeutic options

Comments:

Not Yet Assessable
Not Yet Completed Level 1

Fig. 3.2 Medical Knowledge 2: Therapeutic Knowledge (from *Internal Medicine Milestones*. © Accreditation Council for Graduate Medical Education 2020)

residents work in clinical settings—not just in internal medicine but in all fields of medicine—the work is largely (and for residents, almost exclusively) as members of teams working together in well-defined institutional settings to provide care to delimited groups of patients. Additionally, and crucially important for the purposes of this book, is that the work is

Systems-Based Practice 2: System Navigation for Patient-Centered Care				
Level 1	Level 2	Level 3	Level 4	Level 5
Demonstrates knowledge of care coordination	Coordinates care of patients by effectively engaging interprofessional teams in routine clinical situations	Coordinates care of patients by effectively engaging interprofessional teams in complex clinical situations	Models effective coordination of patient-centered care among different disciplines and specialties	Analyzes the process of care coordination and leads in the design and implementation of improvements
Identifies key elements for safe and effective transitions of care and hand-offs	Performs safe and effective transitions of care/hand-offs in routine clinical situations	Performs safe and effective transitions of care/hand-offs in complex clinical situations	Models and advocates for safe and effective transitions of care/hand-offs within and across health care delivery systems, including outpatient settings	Improves quality of transitions of care within and across health care delivery systems to optimize patient outcomes
Demonstrates knowledge of population and community health needs and disparities	Identifies specific population and community health needs and inequities for the local population	Uses local resources effectively to meet the needs of a patient population and community	Participates in changing and adapting practice to provide for the needs of specific populations	Leads innovations and advocates for populations and communities with health care inequities

Comments:

Not Yet Assessable

Not Yet Completed Level 1

Fig. 3.3 Systems-Based Practice 2: System Navigation for Patient-Centered Care (from *Internal Medicine Milestones*. © Accreditation Council for Graduate Medical Education 2020)

not only goal-directed but, indeed, directed toward shared goals. These goals establish a work (and learning) environment in which residents are constantly discussing and reaching instrumental judgments about "what works." That is, they are constantly making judgments about what ways of thinking and behaving lead to the best outcomes regarding patient care, relations with fellow residents and senior physicians, and interactions with nurses and other health professionals.

The vignette below articulates a senior trainee's goal of using a six-month placement to improve her communication skills.

In supervision sessions, my supervisor and I frequently touched upon my learning goals for the six-month rotation with her and her team. In addition to the technical aspects of the placement, I wanted to improve my communication skills.

My first goal was to learn how to communicate a certain type of diagnosis to patients and their families. I carefully observed how my supervisor managed this task. I was surprised to see that patients and families were able to take the information in, were able to ask thoughtful questions, and did not become distressed or angry. I then practiced this skill myself and found that, over time, it became easier.

My second goal was to learn how to contain anxious or distressed parents when they rang to talk about their children. I watched how my supervisor and other members of the team tackled this task. As I fielded more of the calls, I practiced maintaining my calm, clarifying the issues, taking the unsorted issues back to the team, and following up with the distressed parent. It was never easy, and all interactions were different, but I was increasingly able, over time, to stay composed while containing their anxiety.

The third goal, which emerged from discussions with my supervisor, was to practice establishing and maintaining boundaries (and also to communicate them clearly both to my colleagues and to patients and their families). I practiced leaving work on time, staying late only if required by an emergency. I practiced voicing when my capacity to carry out tasks was reaching saturation. And I was clear when dealing with patients and families that I could do only so much myself; their complying or not with the treatment regime would ultimately determine whether an intervention was successful.

It was only sometime later that I realized the enormity of what I had learned. In a large meeting of doctors and allied health staff, one of my senior colleagues happened to interrupt and correct me multiple times. I was able to contain my agitation and calmly discuss the inappropriateness of her manner with her after the meeting. Because of my composure, she was able to understand her own actions were out of place, and to apologize for them. I was able to navigate a difficult situation and enforce boundaries because of the skills my supervisor took the time to teach me during my placement.

Doing What Comes Naturally

Although we focused above on training in internal medicine and, to a lesser degree, in nursing, much the same analysis would apply to any field of medicine and any other field of health care. The training is demanding and intense, and through study, repetition, and reflective practice[3] (Carroll, 2009; Senediak, 2015), students/trainees transform themselves into doctors, nurses, social workers, clinical psychologists, and various other types of health professional. Slowly, in stages, and through literally thousands of encounters with peers, senior colleagues, patients, and families, trainees reach the point where the skills, behavior, and thinking that may initially have been challenging, difficult, or even obscure or counterintuitive become second nature. Trainees have become part of the interpretive community or community of practice—*insiders* within a community of fellow professionals—that defines and maintains their own professional field, including the diverse range of skills and knowledge that make up that field. As suggested earlier, and as described by Fish in *Doing What Comes Naturally*, they have become "embedded practitioner[s] whose standards of judgment, canons of evidence, or normative measures are extensions" of the community itself (Fish, 1989, p. 144).

In the chapters to follow, we explore how this analysis of "doing what comes naturally" leads to a more clinician-friendly approach to how health professionals understand and address problems in clinical health care and, in the process, in clinical ethics.

Notes

1. Although we could have discussed both medicine and nursing in both the present chapter and Chapter 10, focusing on medicine in this chapter and nursing in Chapter 10 enables us to discuss each chapter's issues in greater depth and with more clarity.

2. The ACGME has also published milestones for the 148 other fields and subfields covered by the Milestones project (MilestonesProject_2). The current list of specialties and subspecialties is available at https://www.acgme.org/what-we-do/accreditation/milestones/milestones-by-specialty/

3. Reflective practice refers to the "capability to reflect critically and systematically on the work-self interface. This process is directed toward fostering personal awareness and resilience" (Gillmer & Marckus, 2003, p. 23; Carroll, 2009; Senediak, 2015). Once a clinical problem has been identified, reflective practice encourages the individual clinician or group of clinicians to create space in which to consider their thoughts and feelings in relation to the problem, as well as the different ways of potentially responding to the problem. In other words, reflective practice engages the whole self in an effort to determine the path forward. By its very nature, reflective practice both engages and promotes informal ethical thinking.

REFERENCES

Accreditation Council for Graduate Medical Education [ACGME]. (2020; implemented July 2021). *Internal medicine milestones.* https://www.acgme.org/globalassets/pdfs/milestones/internalmedicinemilestones.pdf

Accreditation Council for Graduate Medical Education [ACGME]. (2022). *Milestones by specialty.* https://www.acgme.org/milestones/milestones-by-specialty/

Carroll, M. (2009). From mindless to mindful practice: On learning reflection in supervision. *Psychotherapy in Australia, 15*(4), 38–49.

Edgar, L., McLean, S., Hogan, S. O., Hamstra, S., & Holmboe, E. S. (2020). *The Milestones Guidebook. Version 2020.* Accreditation Council for Graduate Medical Education.

Fish, S. E. (1980). *Is there a text in this class? The authority of interpretive communities.* Harvard University Press.

Fish, S. E. (1989). *Doing what comes naturally: Change, rhetoric, and the practice of theory in literary and legal studies.* Duke University Press.

Gillmer, B. & Marckus, R. (2003). Personal professional development in clinical psychology training: Surveying reflective practice. *Clinical Psychology 27*(July): 20–23.

Kleinman, A. (1995). *Writing at the margin: Discourse between anthropology and medicine.* University of California Press.

Lave, J., & Wenger, E. (1991). *Situated learning: Legitimate peripheral participation.* Cambridge University Press.

New South Wales Health. (2022). *Careers as a registered nurse.* https://www.health.nsw.gov.au/nursing/careers/Pages/registered-nurse.aspx

Senediak, C. (2015). Practical guidelines for integrating reflective practice in clinical supervision for psychologists. *Australian Clinical Psychologist, 1*(3), 24–31.

Stanford Medicine. (2024). *Stanford medicine 25: Promoting the culture of bedside medicine.* https://stanfordmedicine25.stanford.edu

Zakowski, L. J., Bennett, N. L., Chheda, S., Esquivel, E., Fazio, S. B., Miller, C. S., & Trowbridge, R. L. (2022). Update and renewal of a national internal medicine curriculum for medical students: Process and outcomes. *American Journal of Medicine, 135*(11), 1382–1386. https://doi.org/10.1016/j.amj med.2022.07.011

PART II

Forging the Clinician's Voice

CHAPTER 4

Building On What's Given

Abstract By the time students enter professional school in health care, they have developed their own distinctive ways of thinking, feeling, acting, and communicating and interacting with others. They have also had long previous experience with roles and role-playing that helps set the stage for their understanding of what is required in their future roles as doctor, nurse, clinical psychologist, or social worker. These assets need to be respected and built upon to yield the clinical understanding and skills for addressing the myriad challenges and problems that arise in daily clinical practice, including those with an ethical component.

Keywords Bioethics · Childhood experience · Clinical ethics · Clinician's voice · Embedded ethics · Ethical capacities · Formal ethical discourse · Health care ethics · History of bioethics · Informal ethical discourse · Preprofessional self · Roles and responsibilities

In the previous chapter we provided an overview of the process by which health trainees develop the skills that progressively bring them to full membership—from outsiders to insiders—in their respective fields. In this chapter and the following two chapters (5 and 6), we set out the foundational ethical capacities that trainees possess, that have been developed over the course of their preprofessional lives, and that are further shaped through their professional training, enabling them to identify, analyze,

S. Scher and K. Kozlowska, *Revitalizing Health Care Ethics*,
https://doi.org/10.1007/978-3-031-78475-0_4

and address the daily challenges, both ethical and non-ethical, of clinical health care. These ethical capacities serve as the foundation for what we refer to as the *clinician's voice*.

IN THE BEGINNING

At the beginning of their professional education in health care, trainees are neophytes in their prospective fields (see Chapter 3), but they already possess the foundational assets that they need in order to develop their capacity to address problems, ethical or otherwise, in clinical health care. That is, by the time they arrive in professional school, they have already had roughly two decades of experience dealing with the ongoing challenges of encountering and living with parents, siblings, relatives, friends, schoolmates, teachers, doctors, coaches, teammates, competitors of various sorts, policemen, workmen, and storekeepers, not to mention myriad brief encounters with strangers in schools, stores, busses, trains, planes, and on the street. They have read novels, watched television, listened to the news, read newspaper stories and magazine articles about current events, and had extended discussions with friends and family about all and any of these. This range of experience is diverse and rich, and it turns each of us into the particular persons we are, which includes the capacity to see ourselves in relation to others and, more broadly, the capacity to perceive and to reason about the ethical dimensions of our local worlds. Included here, too, is the capacity to reason about means and ends in relating to other people and in understanding—and, as necessary, responding to—their needs and concerns.

The following vignette, written by a surgeon, captures the essence and importance of early experiences and their ongoing impact on who we are, lifelong:

> As a preschool child I had a book that proved to be a strong influence on my attitude to life. *Horton Hatches the Egg* [Seuss, 1940] was read and reread many times (in postwar years, books were scarce). Horton, a gentle elephant, promises to look after the bird Mayzie's egg but is subject to many vicissitudes, including the weather, capture, and imprisonment, but throughout he sticks to his promise. As disasters strike he repeats to himself and the world "I meant what I said, and I said what I meant. An elephant's faithful one-hundred percent." Looking back, I realize I have followed this,

trying to avoid promising what I didn't believe possible while always doing my utmost to do what I said I would.

Such formative experiences are common among new trainees in health care. By the time they begin their training, they have all had a diverse range of experiences, including ones that are centrally connected with their determination to enter the health professions. All have developed their distinctive ways of thinking, feeling, and acting that define each trainee as a unique individual and that set the stage for each trainee's characteristic modes of communicating and interacting with others. A friendly and open trainee will, with good fortune, remain at professional maturity a friendly and open clinician. A reticent and circumspect trainee will likely remain at professional maturity a reticent and circumspect clinician, albeit one who has, again with good fortune, identified a satisfying and productive way of meeting the interpersonal challenges of clinical work. We can see this combination of forces at work in the following vignette.

My mother suffered from recurrent and chronic otitis media, which resulted in bilateral perforated eardrums. Although her hearing was impaired, she did not have access to treatment. Communication between us was optimized when we engaged in face-to-face conversation—in the same room, articulating words clearly. We also increased, in time, our use of eye-to-eye language, facial expression, tone, body movement, and so on—all of which improved our capacity to communicate with each other through nonverbal means. That is, in talking with her I also acquired these nonverbal communication skills, and the skills remained with me. Later, after I had become a health professional working in a sexual assault service, I realized I was using these communication skills when conducting assessments of patients. My ability to read body language was particularly important when patients had a history of trauma and could not speak, or when their efforts to communicate about the trauma had been forbidden or disregarded. Also as happened with me and my mother, this process goes both ways. I understand my patients in part through their body language, and my patients comment that my understanding and my underlying feelings are often "spoken" by my facial expressions. All my subsequent training in the field of sexual assault—and in particular, the training about communication skills—has built on the early foundations laid down in communicating with my mother.

We all have a decently concrete idea of the endpoint of training in health care—namely, the emergence of competent, ethically adept, sensitive clinicians licensed to practice, for example, medicine, nursing, or clinical psychology. The crucial question concerns the best way of getting from here—that is, where trainees stand at the outset of their professional education—to there.

A HISTORICAL DETOUR: THE RISE OF (BIO)ETHICAL EXPERTISE

Before addressing the question of how to move forward, a look at the historical roots of modern health care ethics—in very broad strokes—will prove useful.[1]

In the late 1960s, it was becoming increasingly apparent to some observers that advances in modern medicine were creating new, unprecedented ethical problems faster than the medical profession could likely address them. Moreover, the distance between the profession's established practices, ethical and otherwise, and the public's need to address those problems was, if anything, growing. The new field of bioethics soon began to take shape, fueled by the work of the Hastings Center/Institute of Society, Ethics and the Life Sciences, founded in 1969 and located just north of New York City. The Kennedy Institute of Ethics at Georgetown University, another bastion of bioethics, was founded two years later, in 1971.

The emerging bioethics movement was distinctively American in character. The movement's long-standing emphasis on patients' rights and individual autonomy can be tracked back directly to the consumer rights movement of the 1960s, with patients deemed to be the consumers of health care. An early, 1973 article in *Hastings Center Studies*—"Generalization of Expertise" by Robert Veatch (Veatch, 1973)—added an important new dimension to the development of the then-nascent bioethics movement. The article's central thesis was straightforward:

> The problem of generalizing expertise is essentially one of confusing expertise in technical knowledge of a given scientific area with knowledge of what is morally required in that area. (p. 29)

This article had a powerful impact not only on the bioethics movement itself but on medical professionals, who were the experts in the relevant "scientific area." That is, in one fell swoop, the article removed from the medical profession—and by extension, all health professions—any authoritative claim to understanding, or to having the capacity to analyze, ethical problems arising in medicine or health care generally, clinical or otherwise. Included here were the low-level ethical problems that arise in daily clinical practice and that clinicians address daily, on the spot, and, with maturity, almost as a matter of course (Scher & Kozlowska, 2018, 2020) (see also Chapter 1 of this book).

A further implication of Veatch's article was that "ethical expertise" needed to be established; that is, only those with formal training in moral philosophy or bioethics, and not medical doctors or other health professionals, carry the mantle of authority regarding values in medicine and other fields of health care. By the same token, it was only health professionals who embarked on explicit training in bioethics who were considered as having anything authoritative to say about health care ethics; everyone else was a "mere clinician."

The final major element of the formative first decade of the bioethics movement was the publication of *Principles of Biomedical Ethics*, by Thomas Beauchamp and James Childress (Beauchamp & Childress, 1979). The book, published in 1979, is now in its eighth, 2019 edition (Beauchamp & Childress, 2019). The principles in question—autonomy, beneficence, nonmaleficence (better known among health professionals as "do no harm"), and justice, plus respect for persons and truthfulness/honesty—came to be seen as centrally defining elements of the ethical expertise earlier discussed by Veatch. The principles continue to be a dominating force in the bioethics literature, in bioethics consulting, and in teaching ethics (and how to reason ethically) to health care trainees and clinicians.

The current, central role of ethical principles in bioethics teaching is well apparent in the Accreditation Council for Graduate Medical Education's Milestones project, which we discussed in Chapter 3. Taking internal medicine as an example, the only mention of ethics within the set of 21 separate milestones is as an element of professionalism, under "Professionalism 2: Ethical Principles" (p. 22) (ACGME, 2020). A trainee who satisfies the first three levels within this milestone

- Demonstrates knowledge of basic ethical principles
- Applies basic ethical principles to address straightforward ethical situations

– Analyzes complex situations using ethical principles and identifies the
need to seek help in addressing complex ethical situations

What is noteworthy here is the centrality given to ethical principles,
or what we have referred to as *formal ethical discourse*, in defining ethical
competence. Also noteworthy is the apparent assumption that straightforward ethical situations cannot be properly addressed without relying on
ethical principles.

"Mind the Gap"

As we see above, what emerged during this first half century of the
bioethics movement was the view—shared by many bioethicists and also
many clinicians, including clinicians who teach health care ethics to
trainees—that ethical reasoning and ethical action in health care require
adherence to, and application of, ethical principles. That is, they depend
upon what we have described in Chapter 2 as the formal ethical discourse
used by moral philosophers and bioethicists rather than the informal
ethical discourse that is used, literally, by everyone else, ranging from
the man on the street to expert clinicians. As discussed in that chapter,
the philosopher Stuart Hampshire sees informal ethical discourse as
embedded in *ways of life*, which he describes as "coherent totalities of
customs, attitudes, beliefs, institutions, which are interconnected" and
which embody informal "conventions, customs, and practices" (Hampshire, 1983, p. 6). And the social anthropologist Arthur Kleinman
distinguishes the formal ethical discourse—the "abstract knowledge"—of
experts from the informal moral discourse of "social participants in a local
world about what is at stake in everyday experience" (Kleinman, 1995,
p. 45). After noting that "everyday life is inherently a moral process," he
observes that

> the moral is actualized . . . in social transactions over what locally matters,
> often vitally so, such as marriages, family, work, child rearing, education,
> religious practice, health, death. Conflicts among different priorities create
> moral dilemmas as social problems that require action. (Kleinman, 1995,
> p. 45)

In this context what is missing from approaches to teaching and practice that emphasize ethical expertise and formal ethical discourse is a

baseline connection to the lived experience of trainees and clinicians, and to the informal ethical discourse interconnected with that experience.

In Sydney, Australia, riders of the trains are always told, as they are just boarding or exiting a train, to "mind the gap" between the train and the platform. These gaps can be fairly wide since some of the stations are located on curves. We believe that the bioethics movement would have been much more successful, at least from the view of clinicians, if bioethicists had been able to "mind the gap" between what they have perceived as the needs of students, trainees, and clinicians and what students, trainees, and clinicians actually need. This gap was well apparent in our introductory chapter, where the Fletchers needed to make on-the-spot ethical decisions on matters that the hospital's bioethicist saw as requiring potentially a week or more for him to resolve.

What we attempt to do here and in the following chapters is to address this gap by setting out another way of thinking about clinical ethics, at the center of which is (1) the progressive growth and consolidation of trainees' ethical thought, feeling, and action over the course of their training, and (2) the recognition that clinical ethics is not something separate from clinical practice; it is deeply embedded in that practice and in the daily encounters with fellow trainees and clinicians, and also with patients and families. Clinical ethics cannot be understood (or even learned) except in relation to the goals and demands of clinical health care. As Fish notes in discussing the notion of interpretive communities, the values and standards of a professional community are embedded not just in the practices of that community but in the members of the community themselves (Fish, 1980, 1989).

BUILDING ON WHAT'S GIVEN

We opened this chapter by noting that as new professional students begin their education in health care, they bring with them a wide range of human capacities and distinctive ways of thinking, feeling, acting, and communicating and interacting with others. These assets need to be respected, and they should not be squandered. In this context the emphasis needs to be on the *continuity* between the experience and capacities of preprofessional students and their prospective experience and capacities as health care trainees. By stressing the continuity, trainees are in a better position to understand how their health care training will build upon and extend their existing strengths—and not just require the development of new capacities.

What we have in mind, with a view to developing the ethical capacities of health care trainees, is that trainees, in their preprofessional lives, have (1) developed the capacity to appreciate and understand the role and responsibilities that they will have as health professionals, (2) become adept at wielding the same range of ethical concepts that they will be using to understand their professional work, and (3) developed a nuanced capacity for action, including ethical action. These capacities, developed over the course of their preprofessional lives, will stand trainees and (later) clinicians in good stead as they encounter the ethical, social, and institutional challenges presented in their professional work. These capacities are, indeed, the inescapable and necessary foundation for addressing these challenges.

We will discuss roles and responsibilities in the remainder of this chapter, ethical concepts in Chapter 5, and the capacity for action, including ethical action, in Chapter 6.

Roles and Responsibilities

Taking on roles such as doctor, nurse, clinical psychologist, or social worker is obviously, for each trainee, something new. But it's crucial to keep in mind what exactly is new about it. By the time trainees have begun professional school, they have had extensive experience role-playing—experience reaching back to their very early years—and in coming to understand how roles fit into broader modes of organization such as the family, school, team, club, and community.

Just how early is well apparent if we take a look at the website of London's Early Start Group, which includes—under "Nursery" (!)—an illuminating entry: "Importance of Role Playing" (Early Start Group, 2023). In addition to helping "children to make sense of the world around them," role-playing supports children to

Develop their communication and language skills
Act out and make sense of real-life situations
Explore, investigate and experiment
Develop their social skills, as they collaborate with others
Learn to empathise with others, as taking on the role of a character which helps them to learn empathy and understanding of different perspectives

Learn about different cultures
Express their ideas and feelings in a relaxed environment
Develop awareness of themselves and others
Be creative and use their imagination!

Children's games are full of relatively simple role-playing, but as children turn into teenagers and young adults, the roles become more complex and the responsibilities more explicit and more consequential. Being "it" in Duck Duck Goose is a role with responsibilities, but it's nothing like being the leader or other officer of a student organization at the college level. That said, we have all started learning about roles and their associated responsibilities at a very early age, and this learning continues lifelong. And by the time that preprofessional students make the transition to health care trainees, their experience with, and understanding of, various roles is detailed and extensive.

What is worth noting, too, is that as roles become more complex, the need to exercise judgment and to negotiate "next moves" becomes more central. Roles and their responsibilities are intrinsically social, and just what comes next in any situation is often an open question and often a matter of adjustment and negotiation. The range of learning experiences described in the Early Start Group's material on role-playing applies just as well to the learning experiences of adolescents and, later, the young adults who enter careers in health care.

One of the pedagogical challenges in transitioning preprofessional students to health care trainees is to bring along, and build upon, this wealth of experience and knowledge concerning roles and responsibilities. It is, indeed, common for health professionals to have early experiences that serve as the foundation for their later work, as we see in the following vignette.

I am the oldest of three children. My sister was born three years after me, and my brother, the youngest, came seven years later. My mum had trouble conceiving and maintaining a pregnancy. I loved my sister, but the relationship between my brother and me was a gift. It was very special.

I took to co-parenting with enthusiasm as soon as mum was confident that I was safe to be involved. And because I became so skilled in caring for both my siblings, she was able to return to work (she was a nurse) when the family's financial circumstances worsened. No matter what shift or what day my mother was working, I was keen to show my availability and maturity. I remember that my father was not impressed. He was a man of his generation. He was old school. I have a clear memory of my father accusing me of being "a poof" because of my dedication to what he saw as women's work.

My brother was a happy-go-lucky child, attractive, talented, and liked by everyone. He was also sensitive, and he would crawl into my bed when needing comfort (we shared the same room). Getting him ready for school each morning was a delight. Some of the best cuddles from a child are when they have not fully woken. I gave myself time to enjoy those moments.

Not long after starting to attend a Catholic preschool, my brother's behavior changed. He became slow to cooperate in the morning; his mood was low; and he began to cling to me in a way that he had never done before.

On the morning I most remember, he was trembling as I was dressing him. He was clinging to me without speaking. He then proceeded to vomit up what appeared to be all of his breakfast. In helping with my brother and sister, I had learned what it was for a child to look ill, but this situation didn't fit. My brother was not febrile; he did not have any loose stools; and he had slept well overnight. He was pale but not lethargic. There were tears but no respiratory distress.

I cannot say why, but my hunch was that it was the fear of going to preschool that was causing his acute distress. After cleaning him up I told him he would not be going to preschool that day. The response was immediate. His relief was evident in the cuddles and kisses I received.

My mother was something of a lioness when it came to her children. She listened to my description of events and my hunch that something must be going on at preschool to frighten my brother to the point of vomiting. Those days, child abuse was not spoken about. At least not to me.

My mother immediately approached a friend she knew from their earlier nursing days who lived on our street. This wonderful woman immediately said yes to care of my brother while I was at school and my mother was working. I never saw a repeat of my brother's distress, confirming our fears. I don't know whether my mother reported our concerns. Many years later I was told that the nun running the preschool had been dismissed. And many years after that my brother told me that he had been terrorized, struck with a stick, and threatened to be put in a cupboard with the door shut. He was lucky in comparison to others who actually did end up in the cupboard.

As an adolescent, my brother suffered two other traumas over a period of years—one medical and one a serious sporting accident. Some years later, as an adult, his conflict with a boss was one stress too many, and he developed severe mental health issues. Deep down I believed that the emotional abuse that he experienced in preschool had made him vulnerable. I carried this knowledge and burden.

When I grew up and left home, I tried out a number of jobs, but eventually I decided to train as a nurse. I knew from my childhood experiences that I had the instinct for the work, including a combination of skills developed under the tutelage of my mother. My experiences with my brother had developed into a deep-seated concern about the safety and well-being of children. I took all these experiences and skills with me into my nursing career.

The early experiences that serve as the foundation for the trainee's later work can be very textured, as we see in the following vignette.

I grew up on a farm in a large family of six girls. My parents worked hard, and I admired how they strove to provide a safe and nurturing environment for us children. When the weather was good, my three younger sisters and I would be out on the land with my dad, helping with the animals and making hay. Being the eldest, I also saw myself as having a special responsibility. I tried to help as much as I could. I took on a caring role, especially in relation to my younger sisters. My parents relied on me for cooking and cleaning. I was a perfectionist at school and home, setting an example for the others. As I was growing up, my mum often told me I had a calm and caring nature.

From a young age I always wanted to be a nurse; to this day I have never regretted a moment of it. My early life experiences and the relationship I've had with both my parents have carried through in my work as a leader and a colleague. In my family context I learned to be responsible, and I continue to be—as I was from a young age—calm, caring, and confident.

The hospital ward is like my second family. I work with patients from many different cultures and of many different ages. Many of the attributes that I learned and practiced as a child have stood me in good stead when working on the ward. I share food. I encourage others to do well. I form good working relationships, not only with other nurses but with patients, families, and the clinicians on our multidisciplinary team. I look after my staff emotionally—among other things, by listening to them and allowing them to express their feelings. I enjoy working within the amazing family culture that pervades the ward.

Within the health care system, nonclinical staff play a vital role not only in contributing to the patient's health care journey but in helping to prepare patients and families for their clinical encounters with doctors, nurses, clinical psychologists, and other clinicians. Like clinicians, nonclinical staff also come to the job with previous life experiences that shape their capacity to address problems, ethical or otherwise, in their work. The following vignette comes from a medical secretary who coordinated patient care in a busy tertiary-care hospital setting.

Sometime in the early 1980s, I had to attend a maternity clinic for an unplanned (but welcome) first pregnancy. I was determined to enjoy being an expectant mother. For this visit I had written out a list of questions to ask, and I felt very happy and positive.

I sat for a long time in the crowded waiting room. Then, a loud and officious voice announced, "*Miss* T***, Maternity Clinic." It felt like all the eyes in the room looked my way, with sniffs of disapproval at my obvious lack of a wedding band.

On arriving at the counter, the owner of the voice was looking down at me from her high seat behind a glass screen. The sign on her desk said "Manager." She asked who my previous GP was. I told her his name and address, and she said this was not possible. She had worked for that GP, and he did not practice in the suburb in question. She was not to be dissuaded, so I felt obliged to point out this GP had delivered me at home and that I was quite sure my mother remembered the event clearly. After we had argued at length, I suggested she check the phone book, which she then produced with a flourish. It was like a Dickensian scene, with her finger slowly moving down the list—until she paused. There were, in fact, two GPs with the same unusual surname. I thought an apology for her rudeness and appalling attitude would be forthcoming, but alas, no.

After my appointment that day, I didn't have quite the same bounce I had when I started. But new GP at the clinic was lovely, and the midwife and I stayed friends for many years.

The manager's attitude had been so awful, though, that I determined there and then that I wouldn't ever treat another person that way—in such a disrespectful, judgmental, and domineering fashion. And this determination actually served me well in the many years I spent working in the health care system, enjoying throughout a wonderful rapport with patients and their families.

Twenty years later (!), I encountered a woman who had worked at the maternity clinic above. She recalled the day of my first visit very well—and how she and the other receptionists were silently cheering me on as someone who, at last, had stood up to the bully.

Previous life experiences such as the ones above often serve to define for any particular trainee or health care worker what is, in effect, a "template" into which the roles and responsibilities of any particular field need to be "inserted." This point may seem simple and obvious, but much more is at stake than might at first appear. It's not just that being a doctor or a nurse is a new role with new responsibilities; instead, what is at stake in such new roles is that a trainee's past life experience and knowledge can be brought to bear on this role and its responsibilities. Trainees already possess, that is, a well of knowledge and skills for thinking critically and probingly about such new roles and how they fit into broader modes of organization such as the family, school, team, club, or community. In thinking about one's new role as a health care trainee and future clinician, trainees do not come to this challenge as if each of them was a blank slate—*tabula rasa*—where the only questions are "what would a doctor do (or think)?" or "what would a nurse do (or think)?" The question, instead, is "what would *I* do (as a doctor)?" or "what would *I* do (as a nurse)?" In making the transition from preprofessional to trainee, this reliance upon, and relevance of, background experience and knowledge should not be lost, and it should not be dismissed.

What we are suggesting here is that trainees bring forward their own preprofessional experience—their preprofessional selves—as they assume the responsibilities of, and grow within, their new professional roles as, for example, doctors, nurses, clinical psychologists, or social workers. Also important in this context are the preexisting personal resources to be discussed in the following two chapters: trainees' rich understanding of ethical concepts and their nuanced capacity for action, including ethical action. Any failure to bring this range of capacities forward into trainees' professional education would leave them with a minimum of personal resources for learning about and then exercising their clinical and ethical responsibilities.

It is helpful in this context to return to the set of developmental outcomes mentioned by the Early Start Group, including improved language, communication, and social skills, increased empathy and awareness of self and others, and understanding the need for flexible, creative problem-solving. These skills for young children, advanced through role-playing and the recognition of the associated responsibilities, are ones that each person develops over the course of a lifetime. For each trainee, these skills constitute a well of experience and knowledge that they can bring into play as they think, feel, and act in their roles as trainees and clinicians. And for our purposes in this book, the implication is that trainees have extensive experience and knowledge that they can use to address the ethical dimensions of their chosen clinical roles. This experience and knowledge needs to be respected and built upon throughout the course of professional education and training as trainees explore the ethical dimensions of their work.

NOTE

1. In this brief overview of the history of bioethics, we follow the analysis in "The Rise of Bioethics: A Historical Overview," Chapter 3 of *Rethinking Health Care Ethics* (Scher & Kozlowska, 2018).

REFERENCES

Accreditation Council for Graduate Medical Education [ACGME]. (2020; implemented July 2021). *Internal Medicine Milestones.* https://www.acgme.org/globalassets/pdfs/milestones/internalmedicinemilestones.pdf.

Beauchamp, T. L., & Childress, J. F. (2019). *Principles of biomedical ethics* (8th ed.). Oxford University Press.

Beauchamp, T. L., & Childress, J. F. (1979). *Principles of biomedical ethics.* Oxford University Press.

Dr. Seuss (Geisel, T. S.). (1940). *Horton hatches the egg.* Random House.

Early Start Group. (2023, February 2). *Importance of role playing.* https://www.earlystartgroup.com/nursery/importance-of-role-playing/.

Fish, S. E. (1980). *Is there a text in this class? The authority of interpretive communities.* Harvard University Press.

Fish, S. E. (1989). *Doing what comes naturally: Change, rhetoric, and the practice of theory in literary and legal studies.* Duke University Press.

Hampshire, S. (1983). *Morality and conflict.* B. Blackwell.

Kleinman, A. (1995). *Writing at the margin: Discourse between anthropology and medicine.* University of California Press.

Scher, S., & Kozlowska, K. (2018). *Rethinking health care ethics.* Palgrave Macmillan.

Scher, S., & Kozlowska, K. (2020). Teaching ethics in psychiatry: Time to reset. *Harvard Review of Psychiatry, 28*(5), 328–333. https://doi.org/10.1097/HRP.0000000000000258

Veatch, R. M. (1973). Generalization of expertise. *Hastings Center Studies, 1*(2), 29–40.

CHAPTER 5

Dimensions of Moral Experience

Abstract Moral development begins in childhood and continues into adolescence and adulthood, and through each clinician's professional life. Elements of ongoing moral experience—fairness, autonomy, truthfulness, trust, doing good/avoiding harm, solidarity, and moral judgment—are part of the essential background of knowledge, experience, and informal ethical discourse that preprofessionals bring to their health care education and training. These elements of experience represent fundamental building blocks for professional growth, in general, and also, in particular, for what will become the clinician's working moral framework for understanding and addressing the ethical dimensions of clinical experience.

Keywords Autonomy · Clinical ethics · Doing good/avoiding harm · Fairness · Health care ethics · Informal ethical discourse · Moral development · Moral experience · Moral judgment · Solidarity · Trust · Truthfulness · Working moral frameworks

In the last chapter we discussed how trainees' long previous experience with roles and role-playing helps set the stage for their understanding of what is required in such roles as doctor, nurse, clinical psychologist, or social worker. In this chapter we look at other dimensions of trainees' previous moral experience that will come into play in their training and in their later clinical work. We use *moral* rather than *ethical* in this chapter because our primary concern is with the elements of ongoing moral *experience* and how these elements come together as *working moral frameworks*.

THE DIVERSITY AND TEXTURE OF MORAL EXPERIENCE

Theories of moral development deal with overarching stages into which moral experience and thinking are organized. Lawrence Kohlberg's theory, which has been widely discussed and researched, has six stages of moral development, each characterized by a particular type of moral reasoning (Kohlberg, 1969, 1983). The moral notions that define the lowest level of moral reasoning—obedience and punishment—are interpreted and reinterpreted at each progressive stage. Responding to Kohlberg's theory, Carol Gilligan drew on the "different voice" of women to build a theory oriented not toward logic and reasoning but toward human relationships, with an emphasis on caring for, and seeing oneself in relation to, others (Gilligan, 1982). Both theories are concerned, however, with the larger dynamics of moral development. The focus is on *change* over time.

Our interest, instead, is on *continuity* over time. In this chapter we look at a range of moral notions—fairness, autonomy, truthfulness, trust/reliance/respect/reciprocity, doing good/avoiding harm, solidarity, and moral judgment—that emerge early in life, that evolve with age and experience, and that, taken together, set out the dimensions of each individual's moral experience, lifelong. It is these moral notions—coming together as working moral frameworks—that are deeply embedded in the self and that each individual uses extensively in dealing with the life challenges that arise in the normal course of events. By the same token, these moral notions are part of the essential background of knowledge, experience, and informal ethical discourse that preprofessionals bring to their health care training as fundamental building blocks for professional growth.

CONCEPTUAL MAP OF MORAL EXPERIENCE

Fairness

Let's start with the notion of *fairness*. Recent research indicates that by the age of three years, children have developed a sense not just of fairness as such but also of merit and desert—the last two of which would potentially lead to *unequal* allocations without being considered unfair (Blake et al., 2014). Implicit in these findings is that young children are able to determine what it is to be equally situated in relation to a particular potential allocation (e.g., to be equal as children for the purpose of distributing some chocolate) and also what it is to be unequal for the purpose of an allocation (i.e., to deserve different and therefore either better or worse treatment). The following vignette touches upon a child's felt sense of unfairness in the context of a changing family structure:

> My mother divorced my father when I was five years old and then remarried a few years later. I was the only child of her first marriage, and my mother and I were close until she and her second husband had their own children—three additional daughters. In the new family dynamic that developed, my mother did her best to attend to my interests and development, but when push came to shove, my stepfather's preferences and those of his own daughters (not me, his stepdaughter) would carry the day. And in time, even my stepsisters came to see this dynamic as somehow appropriate.
>
> Fortunately, my teachers recognized that I had some academic talent, and I was able, over time, to maintain a strong sense of self and even to win a scholarship for college. The contrast between the school and home environments was, for me, profound. Throughout my adult life, what remained with me was a determination to treat everyone fairly and not to show favorites.

The remarkable thing to keep in mind here is that judgments of equality and inequality, fairness and unfairness, and merited and unmerited will be made literally thousands of times as children are maturing into adults. Family life, school, sports, and even music (e.g., competition for places and then "seats" in an orchestra) all have their comparative, competitive, and allocational elements. And these more or less daily exposures to matters of fairness and equality are punctuated at longer intervals by competitions that will, in effect, determine one's life prospects. Included in this context are determinations as to who is better or best in events ranging over the school, organization, region, state, or country, or who merits acceptance by particular colleges, universities, or professional schools.

Questions of fairness are implicit in all these situations, which may themselves raise important questions of public policy. Witness the long-standing dispute in the United States concerning affirmative action, a policy—recently rejected by the U.S. Supreme Court in *Students for Fair Admissions v. President and Fellows of Harvard College*—that is designed to give some extra preference to those who have been economically or socially disadvantaged. The question of fairness is central, with both sides of the controversy seeing themselves as treated unfairly. Note, too, that affirmative action is controversial not just at the college level but throughout the school years as towns and cities endeavor (or not) to make places in competitive programs or in selective schools available to disadvantaged students.

What the matter of affirmative action makes clear is that overarching social and political frameworks can be characterized as fair or unfair. Questions of social justice and economic distribution, for example, are ultimately matters of fairness, falling along the continuum of fair treatment and fair distributions that all of us become familiar with, and begin to learn about, early in childhood.

What we see here is that judgments of fairness and unfairness can be made from various perspectives. A person might judge one's own actions to be fair or unfair; the same person might judge the actions of another person, whether a family member, teacher, coach, or school or government official, to be fair or unfair; and social and political institutions can also be characterized as fair or unfair, though the terminology used might be that of justice and injustice.

In the following vignette, we see how matters of fairness can be interwoven with the challenges and problems of everyday experience.

During my first year in college, I tried out for the university orchestra. I had just turned down a scholarship (for playing flute) at a well-known school of music, so I was hopeful about the audition. I *was* invited to join the orchestra, but I was disappointed to learn that I would be only *second* chair. Under most circumstances, being offered the position of second chair would be considered an honor for a first-year student, but the situation was not that straightforward. I was confident that I was a significantly better musician than the person positioned as first chair, and the faculty member who led and conducted the orchestra recognized that I was, too.

Instead of simply sending me a pro forma notice about the success of my audition, the conductor invited me to speak with him about the offer. He was very clear that I was definitely the better musician, but he said that the current first chair had had that position for the two previous years. Against that background, the conductor said that demoting the current first chair would be a horrible blow to him but that he (the conductor) also couldn't ask me to play (or, really, to tolerate playing) second chair. He therefore said that he would understand my not wanting to join the orchestra during my first year but that he would surely be in a position to offer me first chair the following year.

I was struck by the generosity and good sense—and fundamental fairness—of this offer. And against that background I decided that the conductor had fully acknowledged my talent and tried to reach a fair, humanly responsive accommodation all around. I realized as much, so I did join the orchestra that year and play second chair.

What's clear in this example (drawn from the college career of a future doctor), and what we need to keep in mind as we consider teaching and learning in health care ethics, is that the notion of fairness/equality/ justice, as well as the other moral concepts discussed in this chapter, is one that is deeply embedded in our life experience and in how we encounter and evaluate our encounters with individuals and institutions. In the example above, the competition for positions in the orchestra obviously raised, in the first instance, matters of fairness and relative merit. But these issues were, in turn, tied in with the expectations of the existing first chair, with the conductor's efforts to nurture the members of orchestra and protect them from harm, and with the conductor's recognition that, on the human level, he had placed the new flautist in a difficult, even painful position. Of special note is that the "solution" to the situation was not via any form of ethical reasoning as such but through the conductor's creative problem-solving and the new flautist's assessment of whether he had been decently treated and with due respect, and whether he could accommodate himself emotionally to playing second chair when he obviously deserved to be first chair. Also, as is typical of many ethical situations in clinical health care, there was no right or wrong answer. The solution reflects the capacities of the persons involved to navigate through the situation and to reach a mutually acceptable accommodation.

Autonomy and Related Concepts

The concept of *autonomy* has a very specific, well-known role regarding patients' decisions in health care, and it also has a distinguished history within philosophy dating back to the moral writings of Immanuel Kant in the late eighteenth and early nineteenth centuries (Kant, 2011, 2015). Those particular, technical interpretations of autonomy, however, are not our concerns here. We are concerned, instead, with a broader notion of autonomy that is closely intertwined with other concepts such as *freedom* and *liberty*, and even more broadly with *privacy, confidentiality*, and *secrets*.

The phenomena relating to this broader notion of autonomy permeate our lives and our interactions with others, day in and day out. The demand and respect for individual independence is a powerful influence on behavior. Efforts to achieve and maintain autonomy color our actions lifelong, beginning in infancy and reaching even into and through senescence. The growth of autonomy in the life of any particular person is interconnected with other elements of the self and is not necessarily straightforward.

> When I was a senior in high school and already accepted to study the next year at a prestigious university, my English teacher told me to stop being so nice. He said that I needed to stand up for my own ideas; if I didn't, I'd be crushed by the competition the next year and thereafter. What I realized then is that my efforts to accommodate to others and to avoid conflict were actually compromising my own intellectual and emotional independence and growth, both short and long term.

Whereas the autonomy of medical decision-making and the autonomy of Kant's moral theory are specifically oriented toward individual decisions and reasoning, respectively, what we see in this vignette is that the broader sense of autonomy is inherently social in character. It concerns the degree and manner in which we see ourselves in relation to others. It also concerns the manner in which other persons or institutions limit, support, or protect our range of action and, by the same token, leave us open, or not, to act in pursuit of our own goals and preferences. Understood in this broader sense, autonomy—concerns about individual freedom and potential reasons, good or bad, for limiting or interfering with it—plays a central role in many relationships, such as those involving families (e.g., between siblings, spouses, and parent and child), schools (e.g., between teacher and student), and careers (e.g., between supervisor and employee). We see this dynamic at work in the following vignette.

> During my elementary school years (through sixth grade), I spent my summers at an overnight camp where we did a lot of camping and canoeing. A few years later, after my first year in high school, a friend (with similar camping experience) and I decided that we would like to take a wilderness camping trip on our own, which would, at the outset, require taking a 16-hour trip on public transportation. Of greater concern to our parents, the trip would involve a full week of camping and canoeing, including some long portages, with no supervision and no way of contacting anyone in case we needed help. Our fathers both refused permission, but they also said that they might decide differently in a couple of years. So, during the summer before our last year in high school, we asked again, and were met with positive, even supportive responses from our fathers. The trip was memorable, but both my friend and I realized that we had gained a certain maturity in the intervening two years, and that that had actually helped us through a couple of tight spots during the trip itself (including in dealing with each other). We both also came to understand how much our *readiness* to take the trip had made it such a success and enabled us to grow from the experience.

In this vignette and in all the settings discussed in the preceding paragraph, one of the central questions concerning autonomy concerns the degree of freedom or responsibility another person can "handle." Virtually any demand for our own greater autonomy, as well as virtually any decision about whether to extend greater autonomy to another, raises questions about the capacity, broadly construed, of the person involved to act effectively within particular constraints or to have a reasonable chance of succeeding at any particular task. And because we are, in our personal and professional relationships, regularly assessing and often pushing against the limits of what we or others can do or should be able to do, assessments of capacity (our own and that of others) are important and recurrent aspects of our lives—individual, social, and institutional/organizational.

Another important aspect of an individual's freedom of action and personal or protected space is the protection of information that is considered confidential, secret, or private. Confidentiality and secrecy are relatively straightforward in that the information is protected by established conventions or institutions (as with confidential information between a lawyer and client or between a health care professional and patient) or by explicit understandings (as when we tell a friend something "in confidence" or "as a secret").

Privacy is more complicated because there are generally no explicit decisions to keep information private (unless, of course, one person specifically directs another to keep the information private). Privacy is, as it were, a moving target. We may tell our spouses or relatives of close friends something that we otherwise take to be private and expect not to be communicated to others. In other cases, information may be more widely known but would generally still not be communicated to one's boss, one's parents, or a large group of coworkers, or to an institution such as a bank, school, or governmental entity. In such cases, "private" information is sometimes better described as "personal" information, but the underlying issue is the same: whether information is considered private or personal, it's just not their business to know. Notably, we make such judgments all the time, but usually just by making ongoing decisions, as we talk, to provide certain information and not other information. In effect, we self-regulate, because we generally sense what to disclose, and what not to disclose, to particular persons about ourselves and about others. Also worth noting is that each of us tends to develop a habitual level of comfort with disclosure about ourselves. Some of us are more private, and therefore less forthcoming, than others.

As with fairness, one's preprofessional experience with autonomy and related concepts is extensive and varied. Likewise, this experience, including the innumerable associated judgments, is ongoing within ourselves as individuals and in relation to other persons and institutions/ organizations. Notably, our impression is that the emergence and explosive growth of the internet and of social media have destabilized what was previously a slowly evolving notion of privacy in Western democracies.

Truthfulness

Truthfulness, when it is once learned as a norm—which is early in life— has an unusual subsequent history. Once we've learned how important it is to tell the truth, we start learning when, for one reason or another, we will not tell the truth, not tell all of it (or any of it), or shade the truth. One type of situation has already been touched upon above. In dealing with private or personal information, we make decisions all the time about what to tell or not tell others. And sometimes the conflicts between truth telling and other moral values or rules can be pretty stark. In philosophy, a conflict frequently discussed in the academic literature concerns what you should do (hypothetically, of course) if, during World War II, a Nazi knocks at your door and demands to know if you are hiding a Jew (which, in fact, you are) (Carson, 2010). Less dramatic but just as conflictual are cases in which you promised someone never to divulge some fact *x*, and then someone asks you directly whether you know anything about *x*. Also of note are situations in which you don't want someone else to know the truth or the whole truth, and in which you therefore try to deflect someone's attention or even to mislead while nevertheless stating nothing but the truth.

Over time, each of us develops skills in dealing with the truth and in determining how much information and how much detail, and at what level of complexity, is needed in order to provide a truthful and, in the context, complete or acceptable answer. We determine what we know; we take account of the setting (family, friends, acquaintances, strangers, schoolwork); we judge what information we want to convey; and we communicate accordingly. The potential complexities and variations are as complex as the diversity of human social interactions. And it would be naive to ignore the phenomena of *lies* and *white lies*, which can themselves vary from the very well-intended (as in the case, discussed above, of lying to a Nazi about the presence of a Jew) to the despicable (as in testifying to the police that you saw someone you despise, but who is completely innocent, commit a murder).

When I was working as a "summer associate" for a well-known law firm, I was given a research/writing assignment by a midlevel attorney. As background, *memorandums* are written for *internal* consumption, to be used in determining the strengths and weakness of a client's legal presentation. *Briefs*, by contrast, are intended for *external* consumption, with the aim of presenting the very strongest legal position in support of one's client. In describing my assignment, the midlevel attorney described two issues that arose in the firm's defense of a particular client. She noted that one of the issues should be analyzed in a memorandum and one as a brief. This struck me as odd, so I immediately asked for confirmation that she intended one of the assignments to be presented as a brief, not in a memorandum (like the other). She reiterated the assignment and said I was correct: one memorandum, one brief. Some weeks later, I was called into the office of the senior attorney in charge of the case and asked why I hadn't highlighted that there was a potentially weak point in our client's position;

the attorney had been surprised when the opposing attorneys presented an argument, with legal support, that wasn't included in my written analysis. I told him that I had been asked to write a brief, not a memorandum, and that that's exactly what I had done. I thought the matter was closed until my exit interview at the end of the summer. The midlevel attorney was present, as was the hiring attorney and also a midlevel attorney who very much wanted me to receive an offer and return the next year. The hiring attorney asked me whether it was possible that I had misunderstood what the midlevel attorney had asked me to do and that I was simply mistaken about being asked to write a brief. The problem was, of course, that I would have had to deny my own very clear experience that I had specifically been asked to write a brief rather than a memorandum (a request that had surprised me) and that the midlevel attorney had reiterated that she wanted a brief (which surprised me again). Although the attorney present who wanted me to return the next year said outright that I should just say "yes, it was possible," I knew that it wasn't. I answered no, and despite the quality of my work throughout the summer, I received no offer (as I expected, given my response). My own take-away from this experience is that the midlevel attorney in question had made a mistake and decided that, for one reason or another, she would attribute the mistake to me. That is, for self-interested reasons, she refused to own the mistake.

What we see in this vignette is the complexity of factors potentially at work in determining whether to tell the truth. The summer associate told the truth because he knew it was the truth and also knew that he would, by doing that, likely preclude his receiving an offer to begin working the following year as a full-time, first-year associate (and at a leading law firm). The midlevel associate either lied or had convinced herself that she was right, likely because she would herself be called to account for her mistake (which did actually cause some embarrassment to the senior lawyer in charge of the case). The midlevel lawyer who wanted the summer associate to receive an offer obviously felt that a white lie was justified in the circumstances. And the hiring partner ended up siding with the midlevel lawyer who had issued the mistaken request—possibly because, to him, the insistence on telling the truth represented an act of defiance, which is not a valued character trait in large, hierarchical law firms.

Telling the truth is not always as easy or as straightforward as it looks.

Trust/Reliance/Respect/Reciprocity

Truthfulness (as discussed above) is tied in, de facto if not conceptually as such, with the interrelated concepts of *trust, reliability, respect,* and *reciprocity.* Truthful communication is the foundation for all four of these phenomena. If a person isn't recognized as a truth teller, then trust, reliance or reliability, (mutual) respect, and reciprocity are all potentially compromised. Note, however, that a person can pick and choose which persons or institutions/organizations to be truthful to and therefore, in turn, which persons or institutions/organizations are ones to which he or she relates with trust, reliance, respect, and reciprocity. For example, an inveterate liar might nevertheless be regularly and reliably truthful to certain family members, friends, institutions, or organizations and consequently be seen as reliable, as worthy of trust and respect, and as capable of reciprocating, in that particular context.

Trust, reliance, respect, and reciprocity are at the very center of our social relationships, including those with family, friends, and colleagues (including, of course, professional colleagues in health care). Social relationships collapse without these features, as do institutions and organizations of all types. More broadly (and as the current, worldwide political situation makes all too clear), democratic or other forms of participatory political institutions equally fail if these qualities are lost.

The following vignette illustrates how a clinician—in this case a family therapist—learned from failures in his own childhood regarding reliance, reciprocity, and autonomy.

My parents were immigrants. Their culture of origin was very hierarchical and oriented toward authority. Respect for, and obedience to, elders was highly valued. Carving out one's own voice and independent sense of self was difficult in this context. At one point in time, when my parents were particularly frustrated by my apparent lack of obedience, they sent me to talk with (and be lectured by) the minister of our church. The minister did not ask me for my point of view or for any explanation as to what may have been the problem. And even if he had, I would not have had the capacity—the emotional resilience—to be able to answer his questions. He simply told me to obey my parents. He quoted Ephesians 6:1–3 (King James Bible, 2017):

> Children, obey your parents in the Lord: for this is right. Honour thy father and mother; (which is the first commandment with promise;) that it may be well with thee, and thou mayest live long on the earth.

Sometime later, during a service, I heard this verse quoted yet again. This time the reading included the final line of the verse, Ephesians 6:4:

> And, ye fathers, provoke not your children to wrath: but bring them up in the nurture and admonition of the Lord.

I remember reflecting at the time that the minister had set out only one part of the truth and that he had failed to implement what would have been, in effect, a family intervention, with the reciprocal responsibilities of *all* parties being brought to light. I carried this experience into my life as a family therapist. It informs my clinical practice to this day. In my daily practice, I work hard to ensure that during family sessions, children and adolescents, as well as their parents, have the opportunity to voice their positions and points of view. I try to make sure that the conversations that ensue include the voice of each family member.

Doing Good/Avoiding Harm

Avoiding harm—initially experienced as pain or discomfort—is one of the most primitive motivations. From our very earliest days, we recoil from anything that harms us. It's part not just of the human condition but of the *sentient* condition.

Also, while we are still very young, we come to understand the notion of harm to others. We are urged not to hit or punch others, not to hurt our pets, and not to hurt or play too rough with our siblings or friends. These are basic lessons in growing up, and we continue—lifelong—to be aware of our potential for harming others. And except for those who have had especially unfortunate childhoods and who obtain some psychological benefit from acting out and hurting other people, we generally try to avoid doing harm to others.

Doing good is more complicated. For some, blessed with kindly spirits, doing good and being generous to others comes naturally, even from a young age. For others, such habits and attitudes develop with maturity. For yet others, these develop incompletely; the importance or moral salience of doing good is recognized only in certain circumstances or in relation to particular persons, organizations, or tasks. And for yet others, doing good and being generous simply fail to connect with the self.

These habits and attitudes with regard to doing good and being generous become further entrenched, though with the possibility for change, in social encounters throughout our lives. These habits and attitudes define our stance toward others and toward the world, and, over time, they become character traits. Also, as with the dimensions of moral experience discussed above, both other individuals and organizations/ institutions can be assessed in terms of avoiding harm and doing good.

Later in the book we discuss the implications for health care training and health care clinicians. For now, two points are worth making. First, it is reasonable to assume that, through their choice of a profession, health care trainees are predominantly persons who are strongly motivated to do good and help others, though it wouldn't necessarily follow that they are therefore generous, too. The two traits are separate. Nevertheless, it is reasonable to assume that, insofar as trainees' or clinicians' professional responsibilities are concerned, the desire to do good (that is, to take care of patients) will define their characteristic approach to their professional work. Within such bounds, generosity or a lack of it are possible attitudes, and no health care professional is necessarily either better or worse because they are not, as a matter of character, generous.

Second, being subject to an obligation to do good or to help others is not the same as acting on any such obligation. Immoral actions of one kind or another are common, of course, but even when people nominally accept or embrace an obligation or moral dictate, they may nevertheless violate it. Organized religions typically include obligations to do good and help others (e.g., consider the Parable of the Good Samaritan within Christianity), but research is at best inconclusive whether religious persons are more likely to do good or to act selflessly or with generosity than agnostics or atheists—who, by definition, do not see themselves as subject to the demands of organized religion. And some recent research suggests that agnostics and atheists, without any formal religious obligation, are likely to act just as well as (or perhaps even better than) theists (Saslow et al., 2013). In any event, feeling subject to a formal obligation, religious or otherwise, is no guarantee that a person will act as the obligation directs. Arguably, what matters most is a person's character and determination to do good and help others.

Solidarity

In the United States, *solidarity* is most often thought of, and invoked, in the context of labor organizations and their role in protecting and advancing the interests of workers, but it also plays a similar role when no specific *labor* organization is involved. The employees in any particular organization, including health care organizations, might feel and promote a sense of solidarity. The notion of solidarity is often invoked on an ad hoc basis for particular causes, as in solidarity with Black Americans, women, teachers, or some other group in the wake of some crisis or crime (e.g., the death of George Floyd [Burch et al., 2021]). And athletic teams, for example, are well-known for promoting and building upon "team spirit," which can be understood as a form of solidarity.

But in Europe, solidarity is a broader, more inclusive notion—a general ethical notion not specifically tied to any specific context (such as one's employment as a particular type of worker, vs. trade-defined unions in the United States). Solidarity is one of the rights included in the Charter of Fundamental Rights of the European Union (European Union, 2000), which defines the rights guaranteed to all persons in EU member states. The Charter's chapter on solidarity includes all the following:

- Workers' right to information and consultation within the under-taking
- Right of collective bargaining and action
- Right of access to placement services
- Protection in the event of unjustified dismissal
- Fair and just working conditions
- Prohibition of child labour and protection of young people at work
- Family and professional life
- Social security and social assistance
- Health care

The intended scope of solidarity is evident in the last three items, which mandate efforts to achieve broad social aims—and throughout the entire European Union, with no limitations regarding applicability.

WORKING MORAL FRAMEWORKS

Bringing together all the above elements is the process of *moral judgment*. Seeing one factor—for example, telling the truth, promoting solidarity, avoiding a harmful consequence, or respecting privacy—as particularly relevant to one's potential course of action is not the same as deciding that one should act in accordance with that one factor. Other moral elements are often in play, as in the case of the Nazi banging at the door and asking if the person is harboring any Jews. Most situations of potential conflict, however, are much less dramatic, as when one needs to balance obligations, responsibilities, rights, potential benefits or harms, and even one's own individual goals in complicated situations. A person might need to consider general questions of work-life balance, including the potentially conflicting interests of any or all of the following: spouse, children, parents, other relatives, and coworkers, not to mention potential others, too. Different people will reach different conclusions, and in many such situations there's no single right answer. It is often easier to determine what *not* to do than to determine exactly *what* to do.

In short, just how the various dimensions of moral experience fit together will vary, potentially dramatically, from one person to another and also, as we have just seen, from culture to culture. We all have distinct personal histories, with the consequence that we have our own individual ways of taking different factors into account, assessing costs, benefits, and risks, balancing one thing against another, and reaching decisions. Each person, in short, has a unique *working moral framework* that we use day to day, minute to minute, in looking at and interrogating our experience, and in interpreting the behavior of others. More broadly, this moral framework is an integral part of the *action framework* discussed in the following chapter, and it has an ongoing, continuous impact on how we think, feel, and act.

Taking a step back, the dimensions of moral experience discussed in this chapter represent some of the most important foundational assets that health care trainees bring to their educational endeavors. These dimensions are continually in play throughout their training, beginning the very first day of their professional education. These dimensions have a fundamental, ongoing role in shaping the natural moral voice of each individual clinician—a voice that needs to be respected, nurtured, and built upon in professional training. More broadly, these dimensions permeate both the informal ethical discourse of clinical health care and the interpretive communities within which that discourse is embedded.

REFERENCES

Blake, P. R., McAuliffe, K., & Warneken, F. (2014). The developmental origins of fairness: The knowledge–behavior gap. *Trends in Cognitive Sciences, 18*(11), 559–561. https://doi.org/10.1016/j.tics.2014.08.003

Burch, A. D. S., Harmon, A., Tavernise, S., & Badger, E. (2021, April 20; updated 23 June 2023). The death of George Floyd reignited a movement. What happens now? *New York Times.* https://www.nytimes.com/2021/04/20/us/george-floyd-protests-police-reform.html

Carson, T. L. (2010). Kant and the absolute prohibition against lying. *Lying and Deception: Theory and Practice* (pp. 67–88). Oxford University Press.

European Union. (2000). Charter of fundamental rights of the European Union. *Official Journal of the European Communities.* https://www.europarl.europa.eu/charter/pdf/text_en.pdf

Gilligan, C. (1982). *In a different voice: Psychological theory and women's development.* Harvard University Press.

Kant, I. (2011). *Groundwork of the metaphysics of morals: A German-English edition.* Cambridge University Press (Original second edition published 1786).

Kant, I. (2015). *Critique of practical reason* (Rev. ed.). Cambridge University Press (Original edition published 1788).

King James Bible. (2017). King James bible online. https://www.kingjamesbibleonline.org/ (Original work published 1769)

Kohlberg, L. (1969). Stage and sequence: The cognitive developmental approach to socialization. In D. A. Goslin (Ed.), *Handbook of socialization theory and research* (pp. 347–480). Rand McNally.

Kohlberg, L. (1983). *Moral stages: A current formulation and a response to critics.* Karger.

Saslow, L. R., Willer, R., Feinberg, M., Piff, P. K., Clark, K., Keltner, D., & Saturn, S. R. (2013). My brother's keeper? Compassion predicts generosity more among less religious individuals. *Social Psychological and Personality Science, 4*(1), 31–38. https://doi.org/10.1177/1948550612444137

CHAPTER 6

Elements of Action

Abstract At the outset of their health care education, all students already possess sophisticated, well-developed action frameworks that enable them to think about ends, means, and the need for ongoing adjustments. These frameworks are supported by capacities and skills that include the following: moral capacities; cognitive, perceptual, and affective capacities; communication skills; theory of mind; quest for narrative coherence; and capacity to recognize our limits. These action frameworks shape, lifelong, the experience and work of clinicians.

Keywords Action frameworks · Clinical ethics · Cognitive, perceptual, and affective capacities · Communication skills · Gut-level feelings · Health care ethics · Homeostatic (physiological) feelings · Narrative coherence · Normal and morally expected course of events · Predictive coding · Recognition of limits · Theory of mind

In this chapter we move from analyzing various substantive dimensions of individual moral experience to examining—in broad, overarching terms—the framework of action that we bring to that experience and that structures how we think about our actions and those of others.

© The Author(s) 2025
S. Scher and K. Kozlowska, *Revitalizing Health Care Ethics*,
https://doi.org/10.1007/978-3-031-78475-0_6

EVERYDAY CONCEPTUAL DISTINCTIONS REGARDING ACTION

Since the conceptual distinctions we make regarding action are so much a part of our everyday thinking about the world, it is easy to overlook how sophisticated and nuanced we are in thinking about our actions and the actions of others.

In the introduction to her 2015 book *Philosophy of Action*, Lilian O'Brien notes the many ways of marking the difference between "my raising my hand when I intentionally wave to a friend, perhaps to signal something to her," and "my hand's rising, where this is the result of a peculiar tic that I suffer from" (p. 2). In particular, the differences between *controlled* action and *uncontrolled* behavior may be elaborated in relation to all the following (for details, see Text Box 6.1):

- Activity—Passivity
- Choice
- Voluntariness
- Purpose
- Reasons
- Trying
- Applicability of a certain kind of why question
- Desire/intention
- Control
- Knowing what
- Knowing how
- Self-evaluation

Text Box 6.1 Everyday Conceptual Distinctions Regarding Action

1. *Activity—Passivity:* The agent is active in the raising, but passive in the rising.
2. *Choice:* The agent chooses the raising, but does not choose the rising.
3. *Voluntariness:* The raising is voluntary, but the rising is non-voluntary or involuntary.

4. *Purpose:* The raising has a purpose, but the rising does not have a purpose.

5. *Reasons:* The agent has a reason to raise her hand and acts on this reason in doing so. The agent does not have a reason to have her hand rise and he doesn't act on any reason when her hand rises.

6. *Trying:* In the raising, the agent is trying to do something, but in the rising, the agent is not trying to do anything.

7. *Applicability of a Certain Kind of Why Question:* A certain kind of why question applies to the raising but not to the rising. Suppose that someone asks of the agent who raises her hand, "Why did you raise your hand?" The question is not inappropriate and the agent may say something like "I wanted to signal to a friend" or "I just felt like it." Now, suppose that the same question is asked of the agent who suffers from the tic. She says, "I didn't raise my hand—it is a tic that I suffer from." The questioner withdraws the question perhaps with some embarrassment. This kind of why question is inapplicable to the rising but applicable to the raising.

8. *Desire/intention:* The agent desires something from or intends something of her raising, but the rising is a movement that occurs independently of the agent's desires and intentions.

9. *Control:* The agent controls the raising, but does not control the rising.

10. *Knowing What:* In raising her hand, the agent knows what she is doing, not on the basis of observing the trajectory of her hand through space, or indeed on the basis of observing anything about herself or her environment; she knows what she is doing in a special way and "from the inside." In the rising, the agent knows what her hand is doing by proprioception and by visual observation of its trajectory through space.

11. *Knowing How:* In the raising, the agent acts on her knowledge of how to raise her hand, but in the rising, the agent does not act on knowledge of how to do something.

12. *Self-evaluation:* In raising her hand, the agent can succeed or fail in the execution of her task and she regards herself as subject to evaluation on this basis. In her hand's rising, the agent does not regard herself as evaluable in this way.

What's of special note here is that these different concepts, all used in exploring the *philosophy* of action, are (1) all themselves simply drawn from our *ordinary experience* and our ways of talking about everyday action, (2) all potentially in play in characterizing particular actions as morally correct or incorrect, and (3) all potentially in play in assessing moral responsibility, praise, or blame.

In clinical ethics, we draw on all of these concepts in determining, for example, how to assess the work of a junior colleague. What was her current level of skills? Did she know what to do and how to do it? Should her seniors have provided better training? What did she actually intend to do? Were there other options available? Could the consequences have been anticipated or avoided? Which elements of the action or situation were in her control and which were not?

All these questions and more are part of how we look at the behavior and actions of other people, whether within or outside a health care setting, whether explicitly moral issues are at stake or not. Importantly, too, these questions are ones we start understanding from the first time our mothers ask "What have you done?" or "Didn't you see what was going to happen?" or when they note that "It's OK. It was just an accident." or "There's no way you could have known." We never stop asking or answering these questions about our actions and those of others. They are, indeed, at the core of many issues in clinical ethics.

THE FRAMEWORK OF PROSPECTIVE ACTION: ENDS, MEANS, AND THE PROCESS OF ONGOING ADJUSTMENT

In thinking about how we are going to act in any particular situation, we all rely on an interactive, dynamic framework to consider the options available and to take action, which may or may not, in any particular situation, have an implicit or explicit moral component. We are continually identifying, though not necessarily on a conscious or verbal level, what we want to achieve and how to achieve it, and we are also continually assessing the potential consequences, risks, benefits, probabilities, and likelihood of success or failure. Second by second, we make adjustments, large and small, in what we are doing and how we do it. And for much of what we do, this all happens so fast that we don't even realize we are doing it.

Salience and Predictive Coding

This continuous feedback loop between our intentions or goals and the world is not simply cognitive or perceptual. What we want and also the feedback we receive from the world inescapably include affective components. One especially important affective component is what neuroscientists refer to as *salience*, which draws our attention to particular elements of the situation at hand, enabling us to focus our perceptual and cognitive resources on what is most important for learning or survival. This affective information is a vital addition to the ongoing feedback loop involved in all human action. More generally, the affective components of action are a crucial element in the process of making real-time adjustments to actions; not uncommonly, before we can articulate that something needs to be adjusted, our feelings are already communicating that something isn't quite what we expected and that we need to make a change, perhaps small, but perhaps also large. We include here *homeostatic (physiological) feelings*, which are often experienced as a "gut sense" that something is wrong or not quite right, that someone is too close to you, or that immediate events are moving too fast to keep properly under control (Pace-Schott et al., 2019). These gut-level feelings, along with predictive coding (see next paragraph), play an important role in identifying problems that need to be addressed, including ones that have an ethical component (see Chapters 7 and 8).

Within neuroscience it is well established that in perceiving the world and acting in it—and, as above, in determining salience—all of us rely on preexisting expectations (for that particular time, place, and context) to interpret what we see. The operative neurological process, which is usually implicit but which sometimes rises to the level of consciousness, is known as *predictive coding* (Barrett, 2017). In perception, the initial determination is whether what one is seeing or hearing, for example, is broadly the *same as*, or *different from*, what was expected. The former allows us to proceed with a determinate range of potential actions (whatever those may be), whereas the latter calls for some sort of adaptive response. Sometimes the response is unconscious and automatic, as when we encounter an unexpected source of danger: our body tightens; our level of arousal increases; our breathing becomes more rapid; our senses grow keener; and we take appropriate action (Kozlowska et al., 2015). Likewise, we recognize immediately when we need to hold something a bit tighter, pay a bit more attention to our walking on an uneven surface or on the beach, or move a bit right or left to avoid someone walking toward us. Sometimes, at the other extreme, we encounter a puzzle or intellectual challenge, requiring that we take a step back and think through something new and previously unencountered in our experience. And we may end up completely changing what we had initially expected to do: the action-in-progress was too risky; the costs and benefits weren't as expected; we changed or misjudged what we wanted to achieve; or the means chosen to achieve the goal or to satisfy the desire were simply inadequate or wide of the mark. Much of our experience, however, lies somewhere between the two extremes, calling for a response that is both neurophysiological, or bottom-up, and cognitive, or top-down—which is what we see in the following vignette.

> I finally bit the bullet and arranged to fix the screen door to my patio. The door would no longer slide. To be opened, the door needed to be pushed, lifted, and pushed again with some force. Sometimes one hand was enough, sometimes not. After the door was fixed, it slid beautifully. A small push with one finger was enough to open the door wide. But at first, it was easy to forget about this change. One time, I tried to open the door with the usual push-lift-push motion, only to see the door fly open and crash, with a bang, against the opposite door jamb. I was so surprised that I almost dropped the plates I was carrying. During the rest of the day, I tried to be mindful when opening the door. It took me some days to adjust to opening the door gently, with only a touch of force. Now that I have readjusted my expectations—or my brain has readjusted its expectations—I no longer have to think about what I am doing. Opening the screen door to the patio is back on automatic pilot.

What we see here is the gradual recognition that a problem existed and needed to be addressed, followed by an effort to fix it and then by a period during which the person's expectations—her predictive coding—gradually moved into line with her silkily sliding door. Here, as in many cases, both within and outside clinical health care, identifying a solution to a problem is one thing, solving it another, and adjusting one's expectations yet another.

The Normal and Morally Expected Course of Events

An intuitive and, for our purposes, especially useful analogue to predictive coding can be found in a widely cited, 1969 article by Harvard's

Robert Nozick, one of the major figures in twentieth-century philosophy (Nozick, 1969).[1] In "Coercion," Nozick starts out by defining a baseline—the *normal and morally expected course of events*—against which one can understand and make judgments about the various modes of influencing other persons' behavior, including not only coercion but threats, warnings, inducements, offers, tips, and rational persuasion. The basic idea in the article is that, for an act to be considered coercive, it must involve a threatened consequence that would leave a person seriously worse off than he or she would expect to be in the normal and morally expected course of events. As with predictive coding, the normal and morally expected course of events encompasses a *range* of potential situations rather than a specific, well-defined state of affairs.

The analysis of coercion is not, of course, our concern here. But the notion of the normal and morally expected course of events is helpful for us in another way. Each person, and each health care trainee or clinician, lives within what is, for him or her, a particular version of the normal and morally expected course of events. Just what is normal and morally expected is a product of an individual's working moral framework (Chapter 5), their own behavior (that is, the individual's present course of action), and how that individual expects "the world" to respond. Included in the latter are both human and nonhuman responses to an individual's behavior. When I walk into a decently upscale clothing store, for example, I expect to be treated courteously and without too much delay. Or when a friend agrees to meet me for dinner, I expect to see him or her there at roughly the designated time or otherwise to receive a decently prompt excuse (or simply an explanation of the delay). And if I am with some friends and announce some especially good news, I expect them to provide some sort of recognition or congratulations. These predictions are based on what we expect of other persons, even when we do not know exactly what they will do in any particular case. By contrast, some expectations are based on the performance of nonhuman systems or even simply objects. If I am old enough to receive Social Security (and if I've properly signed up and contributed), I expect to receive monthly checks. When I put my money into a vending machine for a hot meal (Smart Vending Machines, 2024), I expect to receive the selected meal in return. Or when I pay to go through a turnstile (as for the subway), I expect the turnstile to let me through. These predictions are based on the performance of nonhuman *systems* of various kinds. And if I buy a basketball,

I expect it to bounce when I attempt to dribble it—a prediction about a nonhuman *object*.

And in the normal and morally expected course of events, that's exactly what happens. My friend appears for dinner. I ride the train or receive the vending-machine meal. The basketball bounces when I dribble it. A crucial point here is that the implicit, embedded moral dimensions are typically just taken for granted. We don't even think about them; on a moment-to-moment basis, they're invisible. Even the most complex activities can be understood as instances of the normal and morally expected course of events. Consider the following:

> A few years ago I broke the olecranon of my elbow—the part of the ulna (one of the two bones of the forearm) that provides a cup for the lower end of the humerus (the upper arm bone). I was given an anesthetic that allowed me to remain awake, listen and observe the surgery (or what I could see of it), and even talk with the surgeons. In addition to all the normal cutting and stitching, the surgeons needed to drill a hole in the humerus and insert a pin (which would work its way out months and months later). The surgery was, in its own way, complex, but everything went smoothly, and throughout the operation the two surgeons expressed no particular concern, anxiety, or surprise. They just talked to each other, when necessary, about the surgery, and they occasionally talked with me. Though many things could have gone wrong (starting with the anesthetic and moving on from there, including with the instruments and the drilling), nothing did. Once the surgery was over and I was free to leave recovery, I was picked up by a friend and went home.

And so it is with the normal and morally expected course of events. We arranged to have meet a friend for dinner, so we have dinner together. I paid for the fare, so I ride the train. I bought a basketball, so I shoot baskets with it. These sorts of routine occurrences fill up our days, from sunrise to sunset. We go to work, talk with colleagues, have coffee with friends, encounter strangers on the street or on the train, buy things at the store, discuss the day with our friends or partners, and take care of our children (if any). For the most part, the normal and morally expected course of events simply unfolds. It is our lived experience, shaped by our own intentions and goals. These intentions and goals, however, may or not—as things work out—come to be satisfied. If satisfied, we will likely simply continue on our established course of action. If not satisfied, and as discussed above in relation to predictive coding, we need to make adjustments and set ourselves, if necessary, on a new, revised course of action.

Capacities and Skills That Support the Action Framework

The framework of prospective action, as described above, is something we are all intimately familiar with. It is literally the framework for everything we do. What is easy to overlook, however, is that this single, integrated framework is remarkably complex and comprises many separate elements, many separate capacities and skills. And each person's mix of those capacities and skills—what they are and how they are put together—is unique, defining the particular individual that each of us is. In this section, we fill in the general outline of the action framework provided above.

Of note in this context are the following capacities and skills:

Moral Capacities

The dimensions of moral experience and the working moral framework presented in Chapter 5 are at the center of our action frameworks.

Cognitive, Perceptual, and Affective Capacities

As was apparent in our discussion above of prospective action, our *cognitive, perceptual*, and *affective capacities* are fundamental elements of our action frameworks. These capacities play off one another as we endeavor to understand our experience and its complexities, to identify options, to sort out potential consequences (taking into account the respective probabilities, including likely risks and benefits), and determine exactly what we need to do and how to act in any particular situation. Included here are the *intellectual* and *reflective* skills that enable us to analyze and criticize all dimensions of our experience—what we think, what we see, what we feel, and what we do. Also included here is the capacity to see our actions, skills, and personal characteristics as temporally connected—that is, as the product of *previous* actions, goals, desires, intentions, and plans, and as setting the stage for *prospective* actions or plans. At any particular time, the actions of individuals are part of their own ongoing histories, and both individuals themselves and others are able to understand and interpret actions in this way.

Communication Skills

Our *communication skills* enable us to understand and work with others, to resolve conflicts, and (most importantly for our purposes) to explore and work out the implicit and explicit moral (ethical) dimensions of situations involving other persons, and to reach a working consensus as to what needs to be done. Among the communication skills most relevant to clinical health care are the following: collaborating/cooperating/communicating with patients, peers, clinicians in other fields, and junior and senior clinicians in one's own field; and negotiating with same, in an effort to achieve workable comprises and overcome conflicts.

Theory of Mind

In attributing mental states to others—a core process in defining and understanding our social environments—we rely on what psychologists call *theory of mind*. Beginning in infancy, these skills are learned in tandem with the above communication skills. These skills help us to assess intentionality (i.e., whether some bit of behavior and the consequences thereof were intentional/voluntary/purposeful or accidental), what others know or believe, and whether others are acting honestly and forthrightly (versus, e.g., leaving important matters unsaid or hidden).

Quest for Narrative Coherence

As persons, we have an innate thirst for stories, for narratives. Beginning with fairy tales (= stories with a point), later via books (both fiction and nonfiction), and throughout our lives we endeavor to make sense of our world through stories. Nothing simply happens, whether in the natural world or our human world. There's always a story, and in our efforts to understand why something happened, we're always looking to piece together a set of events or circumstances that hang together, that enable us to understand the narrative linking past and present.

Capacity to Recognize Our Limits

The capacity to *recognize the limits* of what we know or what we can do, and to recognize the need to learn more or to seek assistance, is vital to our well-being and personal growth, whether emotional, intellectual, social, or professional. More specifically for our purposes, this capacity, though surely of ongoing importance in all domains of our lives, is at the foundation of both professional competence and patient safety (Leotsakos et al., 2014) in clinical health care.

THE BRIDGE TO ACTION

Moment to moment, the above capacities and skills are central elements in determining how each of us experiences and defines what is, for each of us, the normal and morally expected course of events. That is, these capacities and skills, coupled with our broader range of beliefs about the social, political, and physical world, determine our moment-to-moment expectations as we encounter the world and prepare ourselves to act.

As with the dimensions of moral experience discussed in Chapter 5, the diverse range of capacities and skills discussed in this chapter are foundational assets that health care trainees bring to their educational endeavors. These dimensions are continually in play throughout their training, beginning the very first day of their professional education. And they continue to be in play throughout their subsequent professional lives as clinicians.

NOTE

1. While a graduate student at Harvard, the first author (SS) had the good fortune of taking courses with Nozick, who remained a friend for decades, until his death in 2002.

REFERENCES

Barrett, L. F. (2017). The theory of constructed emotion: An active inference account of interoception and categorization. *Social Cognitive and Affective Neuroscience, 12*(11), Article 1833. https://doi.org/10.1093/scan/nsx060

Kozlowska, K., Walker, P., McLean, L., & Carrive, P. (2015). Fear and the defense cascade: Clinical implications and management. *Harvard Review of Psychiatry, 23*(4), 263–287. https://doi.org/10.1097/HRP.000000000000 0065

Leotsakos, A., Ardolino, A., Cheung, R., Zheng, H., Barraclough, B., & Walton, M. (2014). Educating future leaders in patient safety. *Journal of*

Multidisciplinary Healthcare, 7, 381–388. https://doi.org/10.2147/JMDH. S53792

Nozick, R. (1969). Coercion. In E. Nagel, S. Morgenbesser, P. Suppes, & M. White (Eds.), *Philosophy, science, and method: Essays in honor of Ernest Nagel* (pp. 440–472). St. Martin's Press.

O'Brien, L. (2015). *Philosophy of action.* Palgrave Macmillan.

Pace-Schott, E. F., Amole, M. C., Aue, T., Balconi, M., Bylsma, L. M., Critchley, H., & VanElzakker, M. B. (2019). Physiological feelings. *Neuroscience and Biobehavioral Reviews, 103*, 267–304. https://doi.org/10.1016/j.neubiorev. 2019.05.002

Smart Vending Machines (2024). Hot meal vending machines. https://www. smartvendingmachines.com.au/hot-meal-vending-machines/

PART III

Moral Engagement in Clinical Health Care

CHAPTER 7

Touchstones for Learning

Abstract The touchstones for learning, which arise at the interface of affect and intellect, mark our encounters with the unexpected. This experience of discrepancy then leads us to ask what has happened, and why? How do we deal with the discrepancy? In this sense our action frameworks, as discussed in Chapter 6, have a fundamental, dual role in clinical experience—first as shaping the background expectations against which incoming experience is assessed as expected or unexpected/discrepant, and second as shaping how we interpret and respond to that experience.

Keywords Action framework · Clinical ethics · Clinician's voice · Discrepancies · Ethical judgment · Expectations · Feelings · Health care ethics · Normal and morally expected course of events · Nurturance · Touchstone process · Touchstone questions · Touchstones for learning · Trainees · Training

A Quick Review: The Foundational Elements of Thought, Feeling, and Action

Before moving ahead, a brief review of our argument in Part II—the three preceding chapters—will prove useful.

Implicit in predictive coding and explicit in the normal and morally expected course of events are the foundational elements of thought,

© The Author(s) 2025
S. Scher and K. Kozlowska, *Revitalizing Health Care Ethics*,
https://doi.org/10.1007/978-3-031-78475-0_7

feeling, and action that influence everything that trainees and clinicians see, feel, and do:

- the existing ethical capacities of trainees and clinicians, including a long-standing awareness of roles and responsibilities, and of how these fit into broader modes of organization such as the family, school, team, club, and community (Chapter 4)
- the specifically ethical dimensions of, and ethical concepts embedded in, our everyday experience, including the following: fairness; autonomy and related concepts; truthfulness; trust/reliance/ respect/reciprocity; doing good/avoiding harm; solidarity; and moral judgment (Chapter 5)
- the action frameworks that we bring to our experience and to everything we see, feel, and do, including the following: cognitive, perceptual, and affective capacities; communication skills; theory of mind; quest for narrative coherence; and capacity to recognize our limits (Chapter 6)

As we discuss in this chapter and the next four, these foundational elements of thought, feeling, and action are continuously working together in four ways. We use them to generate (a) our plans, intentions, and goals, both long and short term, (b) our expectations, via predictive coding, as we encounter the present, (c) our expectations for the future, which include what we expect will happen in the morally expected course of events and also what we expect to happen in relation to our plans, intentions, and goals (i.e., [a]), and (d) our actions in response to what we encounter in the world.

THE INTERFACE OF EXPECTATIONS AND ACTIONS: DISCREPANCIES AND THE TOUCHSTONES FOR LEARNING

In our previous work in health care ethics, we introduced the notion of *touchstones for learning* as a means of identifying ethical issues in clinical health care (Scher & Kozlowska, 2018, 2020). In the context of the present work, we see the touchstones for learning as having three separate dimensions, progressing from

(1) our initial *experience*, via predictive coding and the normal and morally expected course of events, of difference or discrepancy, to

(2) the accompanying *feelings* or "experiences—such as confusion, puzzlement, or surprise—that mark our encounters with the unexpected, unacceptable, or insufficiently understood (and in need of further thought or examination)" (Scher & Kozlowska, 2020) (p. 329)), to

(3) efforts to understand what it is in our current or past course of events that is discrepant or unexpected and that has generated the feelings in (2) above, launching queries along the lines of Text Box 7.1.

Text Box 7.1 Understanding the Discrepant or Unexpected

Ask of oneself or to others:

- Why am I, or why are you, so . . .
- puzzled?
- disappointed?
- concerned?
- angry?
- confused?
- shaken?
- upset?
- challenged?
- surprised?
- delighted

Ask to a group of students, residents, or fellows (as by a supervisor or attending in a small-group conference):

- Has anything happened recently that has left you feeling . . . [same as above]?
- What patients have left you feeling . . . [same as above]?
- Have you had any encounters with other doctors or nurses (or even family members) that left you feeling . . . [same as above]?

Reproduced, with minor adjustments, from "Teaching Ethics in Psychiatry: Time to Reset" by Stephen Scher and Kasia Kozlowska. © Stephen Scher & Kasia Kozlowska 2020

Although Steps 1 to 3 are distinct for analytical purposes—for example, Steps 2 and 3 could not occur without an initial experience of difference or discrepancy—these three steps are, in the real world, often experienced as one, at least on a conscious level. Experienced clinicians, for example, would often just notice that "something's gone wrong here" and then immediately move on to the

(4) *touchstone questions*, which we need to ask in order to understand the clinical situation identified through the touchstones for learning and to determine what needs to be done in order to address, through action, the clinical challenges or problems that it presents. See Text Box 7.2.

Text Box 7.2 Touchstone Questions

- What went wrong, and why?
- What do I/we need to do, in the very short term, to address the resulting deficiency or problem?
- Was there something mistaken or ill-guided concerning my/our execution, planning, or goals?
- Is there something that I/we failed to take into account, that I/we misperceived or misunderstood, or that I/we didn't know?
- Are there potentially competing or conflicting factors that need to be identified and brought into better balance?
- Did I/we misinterpret something that a patient or colleague had said?
- Did I/we misinterpret the intentions or goals of a patient or colleague?
- Is the problem with me/us or with my/our colleague or patient?
- Did my/our colleague or patient misinterpret something that I/we did or said?
- Is there something I/we should have known or asked about, but did not?
- Did I/we, my/our colleague, or the patient somehow overreact to something or otherwise act or react inappropriately?
- Is there something new or unusual about that situation that I/we could not have anticipated?

- Given what happened, what do I/we need to say to our colleague or patient?
- Given what happened, what adjustments need to be made regarding the present situation or clinical goals?
- In the future, how can I/we take steps to avoid a recurrence of what happened in the present situation?
- Is there something about the present situation that should be discussed among a broader set of colleagues and presented as an opportunity for learning or for risk avoidance/management?
- Is the present situation an isolated case or an instance of a broader problem that needs to be identified and addressed?
- What could I/we have done better?
- What are the relevant professional and institutional standards and rules?
- What are the best clinical practices, including as defined in recent systematic reviews?
- Do I/we need to undertake any consultations, if available, with senior colleagues or clinical or outside experts?

© Stephen Scher & Kasia Kozlowska 2025

This four-step progression from the *touchstones for learning* to the *touchstone questions*—which we will sometimes refer to as the *touchstone process*—involves a progression from immediate affective experience (Steps 1 to 3) to the level of clinical problem-solving (Step 4: What went wrong, why, and what needs to be done?). By following this progression, trainees and clinicians identify, and gain a growing capacity to identify, clinical problems at their very earliest stages. In the rest of the book, this touchstone process features prominently in relation to its potential roles both in teaching/training and in clinical practice itself.

See Fig. 7.1 for a more detailed description of the touchstone process, and see the next chapter (8) for a detailed discussion of the touchstone questions.

Baseline experience: The "unexpected, unacceptable, or insufficiently understood" relates directly to our baseline expectations that reflect predictive coding and the normal and morally expected course of events. That is, there is the immediate experience that something is not quite right or quite as expected.

Recognition: What is the particular feeling that this baseline experience produces? This feeling will reflect, in large part, the expectations embedded in the particular course of action we are engaged in, with its attendant thoughts/cognitions/plans that serve to shape the action, our perceptions, and, more broadly, our ongoing experience. The feeling may be a sinking feeling in the gut, a flush of heat, a sudden tension in our jaw, a sense of confusion, a flash of anger, and so on.

Interpretation/Discovery: Why are we feeling that way? This question directly and consciously engages our action framework and, more generally, the full range of foundational elements of thought, feeling, and action as summarized in the text.

Touchstone questions: What went wrong, why, and what needs to be done (regarding past/ongoing actions)? See Text Box 7.2.

Fig. 7.1 The touchstone process: The path from baseline experience to recognition to interpretation/discovery to the touchstone questions. In the real world, the first three steps of the touchstone process—the *touchstones for learning*—are often experienced as one, at least on a conscious level. The three steps of the touchstones for learning are the primary subject matter of this chapter. The *touchstone questions*—the fourth step in this progression—receive only passing attention in this chapter. They are discussed in detail in Chapter 8. © Stephen Scher & Kasia Kozlowska 2025

Two Vignettes

The following two vignettes show how a nurse (first vignette) and a medical resident (second vignette) identify their own distress and use that as the first step in identifying clinical problems and then moving toward constructive clinical action.

La Petite Princesse

In the first vignette, the nurse uses his distress—relating to his perception that the clinical management of a patient was somehow falling short—as a means of understanding his own feeling and thinking, which leads him, in turn, to approach other nurses about his concerns and the problem he had identified. In the longer term, this experience of understanding and acting on his own distress shapes his perceptions of his work and his responsibilities as a senior nurse and team leader.

Juliette was a disabled young infant on the pediatric general medical ward where I worked. Juliette's care included nasogastric feeds and management of hypersalivation while she underwent testing and other procedures in the hope of identifying a plan to maximize outcomes for her. She remained on the ward a long time.

Though Juliette's parents rarely came to visit her, they had not given up their parental rights, and they also remained somewhat protective of their daughter. But it was the infrequency of their visits that drew the particular attention of the nurses who were taking care of Juliette.

Juliette, affectionately referred to as "la petite princesse" (the little princess), had become a favorite of some of the younger nurses. The nurses bought her toys and clothes because they did not like what the volunteers provided. Certain nurses always wanted to be allocated to Juliette or would allocate themselves whenever they had an opportunity to do so. I became aware that Juliettes's mother, on her rare visits, came to sense the possessiveness of the nurses involved in her daughter's care and that she had been, at times, critical of them. Among other things, the mother would have seen her daughter in branded clothes and with toys in her cot that the mother obviously had not bought.

Though I never specifically requested to care for Juliette, I did occasionally take care of her, especially when I was on night shift. Although the night shift was generally busy, we nurses found pockets of time that enabled us to talk while we worked. It was during these conversations on night shift that I was able to give voice to my inner sense of feeling ill at ease. In the process, my concerns about Juliette's management crystallized. I understood that careful, regular attention by a handful of staff—who functioned as substitute attachment figures—was important to Juliette's emotional well-being. But in the case of the branded clothes, I saw this element of the situation—the arguable intrusion of nurses into matters best left to the parents and designated volunteers—as involving a boundary issue (one among others on this ward). As I grew privately critical of some of my colleagues, I also grew critical of myself for my failure to speak up.

When I communicated my concerns during a morning handover, the nurses from the day shift told me that *they* knew Juliette best and were also best positioned to step in for the parents, given their long absences and observable lack of bonding with Juliette. I was also told that Juliettes's plan "was none of my business." It felt like I had been chastised. The team leader did everything in her power to shut down the conversation.

But the conversation did not close down. I was able to talk with a senior permanent night nurse, the team leader for the night shift. I told her about my deepening discomfort: though it was clear to me that the nurses who had become attached to Juliette really cared about her, I also felt that their image of her as the little princess who belonged in branded clothes both objectified her and crossed a professional boundary. Other night nurses joined in this discussion.

While the culture of this particular ward was hard to break, the conversations we had about Juliette, both during night shift and during handovers, represented what was, for me, a major professional change. The conversations were my first experience of recognizing my own distress as the means of identifying an emerging clinical problem (in this case, a boundary problem) and of then, in turn, raising my concerns with my colleagues. After this experience I became more sensitive in noticing boundary issues and also other unaddressed problems. Once I became more senior and began to take on the role of team leader, I was careful how I allocated staff, and I always took boundary issues into account. I also encouraged the members of my team not to keep to themselves any concerns about the quality of care. If something struck them as wrong or inappropriate or even odd—and not just about boundaries—I wanted to hear about it.

What we see in this vignette is a particular nurse's deepening under-standing of the role that feelings such as distress play in identifying clinical problems, in maintaining the quality of patient care, and in signaling opportunities for reflection and professional growth. This is just what we would expect to see in relation to the touchstones for learning and, more generally, the touchstone process.

The Intractably Puzzled Resident

We see a similar dynamic at work in the following vignette (reproduced from Scher & Kozlowska, 2020), though the major source of tension is between doctors and nurses rather than within the nursing staff itself. This difference generates the need for more explicit, more formal processes in efforts to address the clinical situation.

Sophie was a 12-year-old girl with a six-month history of major depression that had gradually developed following her father's sudden death from a heart attack. Sophie had been admitted to the pediatric ward of her local hospital—a teaching hospital affiliated with a nearby medical school—because of poor oral intake of food and fluids. Treatment included intravenous rehydration, followed by a graded increase in Sophie's daily oral intake and daily physical therapy to increase her physical conditioning.

Following the death of her father, Sophie had withdrawn from family life, stopped eating and drinking, and spent most of her time in bed. She self-harmed "as a punishment to myself—kind of a relief that I was being punished. I felt like I needed to be punished." After Sophie's admission to the hospital, the psychiatry resident reviewed Sophie's mental state on a daily basis. Wherever possible, the resident joined the pediatric team on their ward rounds.

Early on, while Sophie was still quite ill, I (a second-year resident in psychiatry) noticed that a senior nurse asked the attending pediatrician when Sophie was going to be discharged. And over the next week, I noticed that the same nurse was raising the same question again and again. The question puzzled me because Sophie remained quite ill and was not even nearly ready to be discharged home. During rounds, when I raised this matter with the attending pediatrician, I was brushed off with the explanation that the ward was always short of beds, that the nursing staff were always asking about discharge dates, and that when push came to shove, the relevant decisions were medical and not for the nurses.

In my weekly session with my supervisor, the supervisor sensed some unusual reticence on my part and therefore asked me whether there was something I had not mentioned. I then explained that my distinct impression was that the ward nurses were, without any apparent reason and contrary to the actual medical situation, trying to get rid of Sophie. The supervisor agreed that the situation was indeed puzzling and asked me what I thought should be done. I suggested that, as a first step, I would discuss the situation with the pediatric residents on the next ward rounds.

The pediatric residents had not noticed anything but agreed that the matter could be addressed at the next group meeting with the attending pediatrician, who responded in roughly the same way as before. But this time, one of the pediatric residents pushed back, saying that there was obviously something going on and likewise reason to suspect that the nurses were somehow distancing themselves from Sophie, with consequent impact on the care and attention she was receiving from the nurses.

After a negotiation with the chief nurse, it was agreed that the nurses involved, along with the chief nurse, would meet with the pediatric residents, attending pediatrician, and me. At the meeting, when the situation was presented to the nurses, they responded that nothing was going on, but the pediatric team refused to accept that as being the full story. And it wasn't.

The nurses then explained that Sophie's mother was intrusive, hostile, impatient, and always dissatisfied with what the nurses were doing; that they simply wanted Sophie and her mother to go home; and, in response to a question from the chief nurse, that they had felt reluctant to raise the matter because they felt it would make them look incapable of coping with a difficult situation.

In the week following the meeting, the chief nurse arranged to have a group meeting on the case in which the main issues were (1) what to do when a close family member is making it difficult for the nursing staff to carry out their normal responsibilities, (2) how to recognize, through their own reactions, that a situation is getting out of control and potentially requires the assistance of senior nurses, (3) the need to recognize that their resistance to answering a question (in this case, from the pediatric residents, about whether something undisclosed was happening) was itself a sign that they needed to think more about the situation itself, and (4) the ongoing importance of their keeping track of their own reactions to their work and colleagues and of voicing their concerns in some appropriate way.

The pediatric team also met with me and my supervisor to sort out what they had learned. A particular focus of this meeting was my refusal to let go of my own puzzlement, my own recognition that something was amiss, and my determination to bring the others' attention to an ongoing problem. There was some tension, however—handled well by my supervisor—concerning the attending pediatrician's initial rejection of the suggestion that something was up with the nurses. The attending acknowledged that he might have been too quick in dismissing my concerns and also that residents were likely well served by both observing, and attending to, the nurses and *their* concerns. My supervisor ended the meeting by noting, with just a tiny bit of irony, "You can learn a lot by listening to yourself."

In this vignette involving the interface between nursing and medicine, it wasn't possible for the young psychiatrist to engage the nursing staff directly regarding his observations and concerns. Given that nursing has its own hierarchy and modes of social control separate from those of medicine, he needed to raise such matters first with his fellow doctors. Largely through his own persistence—and through more formal processes than in the vignette of the young nurses and the baby whom they had called "la petite princesse"—both the nursing and medical staff came to address the problem at hand. Even more importantly in the longer term, through this process the young psychiatrist was able to communicate to both the nurses and his fellow doctors the importance of attending to, and acting on, their own ongoing responses to their work environment—what we have been calling the *touchstone process.*

In these two vignettes, the baseline affective responses of the nurse and medical resident to the clinical situations—the former's distress and the latter's puzzlement—led to their further efforts to understand what was amiss and enabled them to take steps to ameliorate the clinical problems identified. This is precisely the touchstone process as presented above and in Fig. 7.1.

THE DUAL ROLE OF THE ACTION FRAMEWORK

In the above vignettes, the touchstone process led to constructive action, thereby making evident the dual role of the action framework. That is, the action framework, which incorporates the working moral framework, provides the background both for sensing what is "unexpected, unacceptable, or insufficiently understood" and for structuring how to respond to our felt sense of the unexpected. The action framework structures what we are attempting to do in any particular situation, and it is against this background of action-in-progress that we experience something about the outcome as "unexpected, unacceptable, or insufficiently understood." Our action-in-progress determines what is expected or not, and outside the normal and morally expected course of events or not. And the same action-in-progress shapes how to adjust our behavior to the unexpected.

What needs to be stressed here is that within the framework presented in this book and in line with the work of Fish (1980, 1989), Hampshire (1983), and Kleinman (1995, 1999, 2019) (see Chapter 2), our expectations in any situation that involves other persons, as well as our associated perceptions of the unexpected, have an inescapable ethical component. The reason is straightforward. For each of us, our expectations reflect what we see ourselves as doing and what we expect others to do in response. The ethical dimensions of experience discussed in Chapter 5, and as summarized at the outset of this chapter, are a fundamental, though typically implicit, part of our action frameworks. Likewise, the capacities that support the action framework, discussed in Chapter 6 and also summarized above, inescapably shape our actions and perceptions, and what we expect to be able to achieve through our actions.

In our daily actions, much of the above is invisible; the ethical dimensions become visible only when our expectations are disappointed, with the consequence that either our plan of action was somehow not an effective one or that the world or someone in it didn't behave as reasonably expected. As an obvious example, we typically rely on the word of a friend; it is only when the friend has broken his word or told us only part of the truth that we consciously encounter the ethics of everyday encounters. In the clinical setting, the ethical dimensions embedded in clinical care become clearly visible only when the actions or responses of other persons diverge from what we would expect in the normal and morally expected course of events, as when we are uncomfortable or otherwise dissatisfied with what we see in the clinic.

A short way of summarizing the above paragraphs is that we meet experience, and respond to experience, as whole persons. All our capacities are in play as we act: we are, in effect, bundles of capacities and expectations. It is our capacities and expectations that, taken together, lead us to act as we do and lead us to have the expectations that we do. Notably, this action framework also structures our *responses* to the unexpected outcomes of our actions. That is, given that something happened that was unexpected relative to our ongoing action-in-progress, how do we respond? There may be some way of adjusting what we do to recapture or realign the action-in-progress, or it may be that we need to step back a bit, reconsider, and proceed along different lines. In either case, the action framework that we bring to these situations, along with its inherent ethical components, is an inescapable part of how we respond to the unexpected. In reasoning about how to respond, we automatically bring to bear our entire action frameworks, ethics and all.

Consider the following vignette.

I was waiting in line at a restaurant that serves buffet style, with restaurant staff dishing out the main courses. The restaurant was renowned for an especially succulent duck dish. As I approached the head of the line, my anticipation heightened. All I could think was that, within minutes, that glorious duck would be on my plate, ready for me to consume.

But then, when I reached the front of this particular line, the server ignored me and reached out to serve my portion to the person behind me in line. I immediately responded, "Hey! That's not fair!" But the server then said that the person next in line was a close friend of his and that he (the server) preferred to give that portion—the last for the evening!—to someone other than me.

One can sense in this vignette a touch of the gourmand—of someone who really loves to eat.[1] But his single-minded anticipation comes to an abrupt end when the final serving of the dish, contrary to his expectations, goes to someone else. He immediately perceives the situation as *unfair*. This element of ethical judgment is both surprising and unsurprising. The man in our vignette—we'll call him G (for gourmand)—was focused on his coming meal. What does that have to do with ethics or fairness? The answer is: nothing and everything. In the normal and morally expected course of events, as the person next in line, G simply expects to receive the next available portion (which happens to be the last available that day). Instead, the portion goes to someone else, and suddenly the question of fairness, implicit in the situation (being next in line), comes into collision with the facts. Just how our G decides to respond to this unfair treatment is an open question. Will he demand that he be given the portion? Will he complain to the manager? Will he simply leave then and there and vow never to return to this particular restaurant? Different persons will respond in different ways, but what's clear is that the discrepant, unexpected experience of being treated unfairly will be a central determinant.

Note, by the way, that the perception of *unfairness* shapes G's response. Other discrepant experiences would have generated different sets of responses, all of which would have reflected G's expectations and how and why they were disappointed in any particular case. If the server had, in serving G, somehow managed to drop the duck onto the floor (another discrepant experience), G would surely have been disappointed but not seen the situation as involving an ethical failure. Likewise if the server suddenly realized that the manager had previously told him to put aside a portion for a long-standing customer coming in that night. The server might have withheld the portion, and G might have been sorely disappointed. But again, it's not clear that the response would have had an ethical component. This response would subsequently change, of course, if the server took the portion into the kitchen, told a busboy to take it out to his friend after a few minutes' delay, and G noticed—and realized that the server had lied to him. As with the server's initially unfair treatment (our original example), just how to respond to the server's obvious lie would vary from person to person.

If we replace the above scenarios with any exchange between a health care clinician and another clinician, a patient, or a family, the above analysis would likewise apply. Normal communication is normal communication. If it goes as expected (with the normal and morally expected course of events construed broadly; see, e.g., Chapter 8 and the vignette with Dr. Sassall in Chapter 10), it's done and one moves on. But in response to discrepant experience, one needs to determine the origin of the failure or discrepancy. Why has it happened, and once that question is decided, how should one respond? Has the clinician done something incorrectly or done something that led to a discrepant outcome? Or is there something discrepant about the patient's response, such as their resistance, evasion, or outright refusal? The clinician needs to settle such questions in order to determine how to proceed—that is, whether to continue pursuing the previous clinical goal or to change the goal, with the consequence the clinician settles for second best and tries to determine how else, under the circumstances, to help this particular patient.

It is not always the trainee or clinician who recognizes the discrepancy, as we see in the following vignette.

An 80-year-old patient experienced a complication of surgery for insertion of a pacemaker. Her ventricle had been pierced, and she had developed a pericardial effusion. Following the insertion of a new pacemaker and the successful drainage of the effusion, she was prescribed colchicine to settle any pericardial inflammation. But when checking the medications dispensed by the hospital pharmacy, she noticed that the fact sheet for the colchicine was headed "Colchicine for gout attacks." Her immediate response was, "But I don't have gout." The pharmacist, after explaining that the fact sheet was the only one that they had pertaining to colchicine, verbally supplemented the information in the fact sheet with what the patient needed to understand regarding its use for pericardial inflammation.

In this vignette the pharmacist's response to the discrepancy was straightforward, but as we discuss in the following section, trainees and clinicians do not necessarily respond appropriately, thereby raising questions about what needs to be done in order to maintain and build upon the human assets (Chapter 4 to 6) that serve as the foundation for the clinician's voice.

NURTURING FOUNDATIONAL ASSETS/ THE CLINICIAN'S VOICE

Consider the following vignette by a senior physician, in which a medical resident fails to own a mistake.

The patient, an adolescent boy, had presented with seizures and behavioral disturbance. His brain scan was normal, and results from the lumber puncture were pending. In the meantime, the neurology and psychiatry teams had settled on a sedation plan. Of particular relevance, the patient's medication would be administered in wafer rather than tablet form because he was considered too unwell and likely too uncooperative to swallow a tablet. The pediatric resident was given the responsibility for entering the sedation plan onto the patient's medical chart. When I came to the patient's bedside on afternoon ward rounds—the second round of the day—the patient's mother, who had not left the bedside, began to cry in distress. Between sobs, she told me her son had been in a state of fearful screaming for more than an hour earlier that day because he had been unable to take his prescribed medication. What became clear is that the pediatric resident had mistakenly specified tablets rather than wafers, that the patient (as expected) was unable to take the tablet, and that he deteriorated over the course of the hour that it took to correct the medication order in the chart. When apprised of the situation, I apologized for the error, and I looked at the resident, leaving her space to apologize also. There was silence. When we had left the room, the ward's charge nurse approached me and complained that the resident had blamed one of her nurses for not dispensing the wafer form of the medication. The charge nurse emphasized that her nurse had dispensed exactly what had been specified in the chart— the tablet. I then turned to the resident, saying that we, the attending physicians, had seemingly not been sufficiently clear that we had wanted the patient to receive the wafer formulation, and I apologized if that was the case. I also suggested that, whatever the background logistics, the resident needed to apologize to the mother. The error had caused significant distress, and an apology was needed. The resident needed to repair the relationship as we would be looking after this patient for some time to come. The resident looked shifty. When I checked later in the day, the apology had still not happened. Also not the next day. It was then that I knew that we had a problem.

In the above vignette, the resident was expected to admit her mistake and apologize (as suggested) to the mother. But that's not what happened. The resident attempted to blame a nurse and subsequently failed to apologize on the spot, or the next day, or the next, in response to which one can see the deepening concern of the senior physician:

Every day I was hopeful that the resident would step forward, own her mistake, and make amends with the mother and nurses. And every day I found myself discarding one explanation after another for the resident's failure and, along with it, yet another idea of how I would then follow up to support the resident in finally confronting the realities of the situation. By the end, it was clear that the resident's ongoing failure needed more attention than I had initially expected. In my weekly supervision session with the resident, the entire hour was taken up, as happens occasionally, with *reflective practice*.[2] My goal was to create a place for the resident to consider her thoughts and feelings in relation to the current clinical problem, as well as the different ways of potentially responding more constructively to the problem. In particular, why had the resident found it so difficult to own the mistake and make the requisite apology? The further hope, of course, was that the resident would, in the future, be more in contact with her own thoughts and feelings, and also be more able to deal more constructively with the unexpected—regarding both patients and other professionals.

In this vignette we see what looks to be a failed effort to enable the medical resident to look more closely at herself, her work, and her dealings with patients and other clinicians. From the perspective of the current chapter, the resident has failed to engage with the touchstones for learning—the questions that would set her on the path toward addressing the various problems described in the vignette.

The question for us is the following: how or why has this happened? Presumably, at the outset of her medical education, the resident possessed, even to an outstanding degree, the foundational assets that we have described as the basis for forging the clinician's voice: an understanding of one's roles and responsibilities; the multidimensional moral experience, along with one's working moral framework, that has been developing lifelong; and the action framework that one brings to everything one sees, feels, and does. Something has obviously happened to this resident that has prevented her from bringing this experience and these capacities to bear on her present work. But rather than hypothesizing what may or may not have gone wrong in the life of this one resident, what the vignette highlights is that the foundational assets in question need to be maintained and nurtured over the course of professional education and clinical training. Research suggests that the failure to nurture foundational assets during the training process results in the erosion of baseline human responses such as empathy and compassion (see Chapter 12 in Hojat [2016]). The quality of patient care ultimately depends on clinicians' capacities to bring into play the full scope of their human capacities— what we have called foundational assets—as they endeavor to care for their patients (Hojat, 2016; Kleinman, 2019).

In our view, one way of maintaining and supporting these capacities— and of nurturing the clinician's voice—is for teachers and clinicians to continually draw attention to the touchstones for learning and, more generally, the touchstone process, and to encourage students, trainees, and other clinicians to keep this process in mind as they engage in their clinical work. Given that the touchstones for learning are situated at the interface of feeling and cognition—as in "What just happened?" "Why are you so puzzled?" "That makes no sense!"—these encounters with the unexpected invite and support both an openness to feelings and an invitation to understand those same feelings, both within oneself and in relation to the broader socio-institutional environment. As such, the questions support the growth and maintenance of both the human capacities underlying the provision of quality health care and the cognitive capacities for identifying and addressing the (often-embedded) ethical challenges that arise within clinical care.

NOTES

1. As we were polishing the final draft of this book, we ran into the following in the *New York Times*—in an article in praise of the plain croissant (Rao, 2024). Waiting in line at a favorite bakery, the author observed:

It was a delight to see people in line—the plain croissant still had its fans! But as I studied the small crowd and did the math, it was also a little alarming: What if there wasn't one left for me by the time I got inside?

2. See, for example, Carroll (2009), Gillmer (2003), and Senediak (2015).

REFERENCES

Carroll, M. (2009). From mindless to mindful practice: On learning reflection in supervision. *Psychotherapy in Australia, 15,* 38–49.

Fish, S. E. (1980). *Is there a text in this class? The authority of interpretive communities.* Harvard University Press.

Fish, S. E. (1989). *Doing what comes naturally: Change, rhetoric, and the practice of theory in literary and legal studies.* Duke University Press.

Gillmer, B. & Marckus, R. (2003, July). Personal professional development in clinical psychology training: Surveying reflective practice. *Clinical Psychology 27,* 20–23.

Hampshire, S. (1983). *Morality and conflict.* B. Blackwell.

Hojat, M. (2016). *Empathy in health professions education and patient care.* Springer International Publishing.

Kleinman, A. (1995). *Writing at the margin: Discourse between anthropology and medicine.* University of California Press.

Kleinman, A. (1999). Experience and its moral modes: Culture, human conditions, and disorder [lecture delivered in 1998]. In G. B. Peterson (Ed.), *The Tanner lectures on human* values (Vol. 20, pp. 355–420). University of Utah Press; Cambridge University Press. https://tannerlectures.org/lectures/experience-and-its-moral-modes-culture-human-conditions-and-disorder/

Kleinman, A. (2019). *The soul of care: The moral education of a husband and a doctor.* Viking.

Rao, T. (2024, July 17). The plain croissant is still the best croissant. *New York Times.* https://www.nytimes.com/2024/07/17/dining/plain-croissant-best.html

Scher, S., & Kozlowska, K. (2018). *Rethinking health care ethics.* Palgrave Macmillan.

Scher, S., & Kozlowska, K. (2020). Teaching ethics in psychiatry: Time to reset. *Harvard Review of Psychiatry, 28*(5), 328–333. https://doi.org/10.1097/HRP.0000000000000258

Senediak, C. (2015). Practical guidelines for integrating reflective practice in clinical supervision for psychologists. *Australian Clinical Psychologist*, *3*, 24–31.

CHAPTER 8

Informal Ethical Discourse and the Touchstone Questions

Abstract The *touchstone questions*—the last step of the *touchstone process*—play a central role in informal ethical discourse, clinical decision-making, and health professionals' communication with other health professionals and with patients and families. As we will see, the touchstone questions engage the whole person, and it is as whole persons that clinicians confront the ethical challenges and problems presented by their clinical work.

Keywords Action framework · Between the flags · Clinical ethics · Health care ethics · Informal ethical discourse · Normal and morally expected course of events · Moral judgment · Touchstone process · Touchstone questions · Touchstones for learning

Our discussion of the dimensions of moral experience and working moral framework (Chapter 5), action framework (Chapter 6), and touchstones for learning and touchstone process (Chapter 7) enables us to extend our analysis of informal ethical discourse and of what it is for such discourse to be embedded in clinical health care. In *Rethinking Health Care Ethics* (2018), and in line with the discussion by John Rawls in *A Theory of Justice*, his landmark 1971 work on political philosophy, we considered informal ethical discourse to be the ethical language used by the proverbial man on the street—that is, by all of us, including health care clinicians

(Rawls, 1971; Scher & Kozlowska, 2018). In Chapter 2 of *this* book, we saw how informal ethical discourse is embedded in Fish's interpretive communities (Fish, 1980, 1989), Hampshire's ways of life (Hampshire, 1983), and Kleinman's moral modes of experience (Kleinman, 1995, 1999).

But when we look at informal ethical discourse in the context of clinical health care and the action framework that clinicians bring to their work, we see something new and important. In this clinical context the action framework we discussed in Chapter 6, along with its inherent ethical dimensions as discussed in Chapter 5, serves to structure and express how health care clinicians think, feel, and act—and therefore how they use informal ethical discourse within a clinical setting.

When situations fall within what was expected by predictive coding—when they fall within the normal and morally expected course of events—our responses to them are typically straightforward. Examples from the last chapter include meeting, as planned, a friend for dinner; receiving, as expected, congratulations for something special that we have achieved; and actually being served the portion of duck that we had been waiting in line to receive. Examples of projected outcomes drawn from the clinical context include myriad procedures and processes such as vaccination, catheterization, intubation, safety measures, prescription and administration of medications, outpatient (day) and inpatient surgeries, and clinical, diagnostic, and risk assessments. If properly conducted, the outcomes of these procedures and processes are likely to fall within a predictable range—the normal and morally expected course of events.

But sometimes the situations we confront fall at the edge of, or clearly outside, the normal and morally expected course of events. To use the language of predictive coding, we realize that something *salient* is *different*. In ordinary language, we experience a discrepancy between what we were expecting and what we actually perceive or encounter. The same goes for clinicians in clinical practice. In such situations, the ethical issues implicit or explicit in the situation may require only very slight, low-level, possibly even unconscious adjustments in the clinician's thoughts, feelings, or actions. For example, if a family has experienced significant past trauma, the clinician, when taking a clinical history, will take particular care to ask questions in a way that is sensitive and supportive. But sometimes the situation is less straightforward, and the clinician may need to consider the situation more directly. Doing so may take only a few quick moments of thought; or it may require the careful consideration of the options, consequences, and human relationships involved; or it may even be that the clinician cannot move forward without discussing the matter with another clinician, the department head, or even a senior administrator. In some unusual or extreme cases, the clinician—or clinical team—might even decide to meet with someone having particular expertise (including a bioethicist, lawyer, or cultural consultant) or with a trusted member of the clergy who might bring some special dimension of wisdom to the situation.

THE TOUCHSTONE QUESTIONS AND THE INFORMAL ETHICAL DISCOURSE OF PATIENT CARE

The Touchstone Questions

When the first author (SS, years ago) taught ethics clinically in Harvard-affiliated hospitals, he encountered an oddity about how people stored "ethical" information. If he asked doctors, individually or in groups, whether there were any ethical problems that they were worried about, the answer was almost universally *no*. But if he asked doctors (or other clinicians) whether there were any cases or situations that were *bothering* them—situations that diverged from what they were expecting or hoping to occur—the answer would be an immediate and resounding *yes*. For example: Mr. Johnson insisted that he was suffering from the same disease as a recently deceased friend and was demanding medications that were unrelated to his actual illness; Mrs. Jones and her family were in regular conflict with the medical staff, including the nurses, because they expected that the response to treatment would be immediate, instantaneous; the chief resident was preoccupied with some matters at home and was losing track of his responsibilities at the hospital; and the nursing staff was finding it stressful to work with a newly recruited doctor who was pushing innovations in treatment and patient management that ran against the hospital's established practices.

All the above situations involve human, interpersonal, or institutional components that are not open to technical, determinate solutions. The situations all raise what are, in effect, ethical issues, but the issues involved are generally not identified (or, in practice, analyzed) as such. In a clinical setting, they're just "problems" that arise against the background of what we were attempting to achieve or, more broadly speaking, what we were doing. And in trying to address these problems—the last step in the *touchstone process* described in Chapter 7—clinicians ask the same sorts of questions that they would use in addressing any technical, medical, or scientific problems, lapses, or failures. Since these questions arise as the final step in response to the touchstones for learning, we refer to them as the *touchstone questions*. See Text Box 8.1.

Text Box 8.1* Touchstone Questions

- What went wrong, and why?
- What do I/we need to do, in the very short term, to address the resulting deficiency or problem?
- Was there something mistaken or ill-guided concerning my/our execution, planning, or goals?
- Is there something that I/we failed to take into account, that I/we misperceived or misunderstood, or that I/we didn't know?
- Are there potentially competing or conflicting factors that need to be identified and brought into better balance?
- Did I/we misinterpret something that a patient or colleague had said?
- Did I/we misinterpret the intentions or goals of a patient or colleague?
- Is the problem with me/us or with my/our colleague or patient?
- Did my/our colleague or patient misinterpret something that I/we did or said?
- Is there something I/we should have known or asked about, but did not?
- Did I/we, my/our colleague, or the patient somehow overreact to something or otherwise act or react inappropriately?
- Is there something new or unusual about that situation that I/we could not have anticipated?
- Given what happened, what do I/we need to say to our colleague or patient?
- Given what happened, what adjustments need to be made regarding the present situation or clinical goals?
- In the future, how can I/we take steps to avoid a recurrence of what happened in the present situation?
- Is there something about the present situation that should be discussed among a broader set of colleagues and presented as an opportunity for learning or for risk avoidance/management?

– Is the present situation an isolated case or is an instance of a broader problem that needs to be identified and addressed?
– What could I/we have done better?
– What are the relevant professional and institutional standards and rules?
– What are the best clinical practices, including as defined in recent systematic reviews?
– Do I/we need to undertake any consultations, if available, with senior colleagues or clinical or outside experts?

* © Stephen Scher & Kasia Kozlowska 2025. This text box, identical to Text Box 7.2, is reproduced here for the convenience of the reader.

These touchstone questions are central to the informal ethical discourse regarding the care of patients. That said, they are noteworthy in several respects: none of them is explicitly ethical; none of them uses ethical language; and the outcome or discrepancy is assessed specifically in relation to our own clinical intentions and goals. In asking these touchstone questions regarding past actions and ongoing actions, we are looking at the clinical situation pragmatically, largely in terms of ends, means, potential misunderstandings and misperceptions, other possible failures of various kinds, and the need for adjustments and improvements. In a nutshell, in our effort to understand the discrepancy between our intentions/goals and the outcomes of our actions, we are, in effect, using the touchstone process (Chapter 7) as a means of bringing forward in time our action frameworks and their supporting capacities (Chapter 6), including the dimensions of moral experience (e.g., fairness, autonomy, and truthfulness) that make up our working moral frameworks (Chapter 5). It's important to note, too, that these working moral frameworks include and bring into play the full range of ethical concepts (e.g., rights, responsibilities, obligations, permissions, prohibitions) that are part of our everyday, informal ethical discourse. That said, although these concepts are essential parts of our informal ethical discourse, we often do not use specifically ethical terminology in referring to them. "I have an obligation to" becomes "I need to"; "It is permissible for you to do x" becomes "It's up to you whether you do x"; and so on. For more on such equivalences, see Chapter 9.

As we consider how to move ahead in any particular clinical situation, the implicit ethical issues of informal ethical discourse are intermingled with issues involving technical skills and tasks. This identification of "problems"—and consequently of implicitly "ethical" problems—takes place against the ongoing background of the normal and morally expected course of events. That is, the background condition is that the persons are acting, or that the institutions are structured, in ways that are consistent with what is normal and morally expected in that setting. It's not that the ethical dimension is present only when something is wrong or diverges from the normal and morally expected course of events. The ethical dimension is *always* present, and it's specifically because the situation diverges from what is normal and morally expected that one can identify what are, in effect, ethical problems when they do arise.

The Fundamental Role of Moral Judgment

The process of asking and answering the touchstone questions generates new information, and potentially quite a lot of new information. But being in the possession of new information does not, in itself, generate a conclusion about how to respond or what to do. Doing so is a matter of judgment, and since ethics is interwoven with everything clinical, the judgment involved is, in part, moral judgment.

Judgments made from particular collections of information will inescapably differ, sometimes dramatically, from person to person. The foundational elements of the clinician's voice that we discussed in Chapters 4 to 6 come into play here. Students, trainees, and clinicians continually develop

(1) the capacity to appreciate and understand their role and responsibilities as health professionals (Chapter 4),
(2) their understanding of all the following in relation to their work as health care clinicians: fairness; autonomy and related concepts; truthfulness; trust/reliance/respect/reciprocity; doing good/avoiding harm; and solidarity. All these are brought together through moral judgment (Chapter 5), and
(3) their cognitive, perceptual, and affective capacities; communication skills; quest for narrative coherence; theory of mind; and capacity to recognize their limits (Chapter 6).

But just how these many elements actually develop over the lifetime of any particular clinician, and just how different clinicians balance them one against another, and at any particular time, is subject to wide variation, encompassing the full range of human psychological growth.

Recurrent questions, to be answered by different trainees and clinicians in different ways, include the following:

- In dealing with patients, families, and colleagues, just where do I draw the boundary between what I deem acceptable or necessary and what is not?
- How fair is fair?
- In respecting or promoting the autonomy of others, when have I done enough? And how hard should I push in pursuing my own interests and goals?
- How much information do I need to convey, and in what form, in telling others the truth? Likewise in judging whether others have been sufficiently truthful or forthcoming.
- Are my expectations realistic in judging the reliability of others or whether they have acted with due reciprocity?
- When do I say I've done enough to promote others' good or to avoid harm?
- How does our solidarity, or lack of it, with others (especially colleagues) influence our actions and expectations, and how much do they or I assume can be taken for granted?
- In communicating with others, how much clarity can I expect of others, or of myself? How much do I expect them or myself to make everything explicit?
- In looking to make sense of a story—or my environment, including professional setting—how much coherence do I want to see? How far do I pursue loose ends or missing details?
- In the context of group conflict or collective decision-making, how flexible should I be, and how flexible should I expect others to be?
- My limits are bounded but flexible. How hard do I push myself, and when and how?

It is as distinctive individuals that each student, trainee, and clinician will learn, develop, and mature as a health care professional. The overall goals for training may be the same within a particular field of health care, but trainees will each have their own particular ways of achieving and understanding those goals, and of integrating them long term into their own professional styles. These elements of judgment will come into play whenever health professionals pose, and try to answer for themselves, the touchstone questions. They will also come into play whenever health professionals engage in various sorts of group processes, including efforts to resolve conflict with colleagues, administrators, or patients and their families.

BETWEEN THE FLAGS

An example drawn from the Australian "Between the Flags" program for clinical monitoring will bring into focus just what it is for the normal and morally expected course of events to serve as a baseline for identifying clinical problems, ethical or otherwise, that need to be addressed.

Cardiac arrests are common in-hospital events associated with high mortality (Andersen et al., 2019). The majority of arrests are preceded by physiological instability that manifests as deteriorating vital signs—ones that have shifted outside of the homeostatic zone, the parameters needed to sustain human life. These clinical situations require quick intervention if the patient is to survive.

In the Australian state of New South Wales (NSW), the death of a 16-year-old girl in 2008 prompted a formal inquiry and spurred the NSW Clinical Excellence Commission to take action. The Commission assembled a team of physicians, surgeons, nurses, and other clinicians for the purpose of developing a standardized rapid-response system called "Between the Flags." The program draws on imagery from Australian beaches, where volunteer lifesavers put up flags—red above and yellow below—between which it is safe to swim, without fear of dangerous currents. In Australia, all beachgoers know that to stay safe, they need to swim *between the flags*, where the lifesavers monitor and are positioned to provide swimmers with assistance if they get into difficulties. Using the color theme of the flags, the Between the Flags program for hospitals mandates that clinical charts include three zones: white, yellow, and red. "Observations 'between the flags'"—that is, in the white zone between the yellow zones in Fig. 8.1—"indicate a stable and safe patient" (Hughes et al., 2014, p. 715). All is going well, and no immediate action is required. Observations outside the flags and in the *yellow* zones require that the patient be "monitored and seen by someone from the admitting or home team within half an hour if the nurse in charge of the ward considers this necessary" (Hughes et al., 2014, p. 715). Observations outside the flags and in the *red* zones reflect dangerous levels of vital signs and the need for immediate action. "Whenever a nurse enters a reading in the red zone, an immediate referral must be made to the RRS [rapid-response system] team" (Hughes et al., 2014, p. 715). The Between the Flags program has been implemented in all NSW hospitals from 2010. It is "associated with continued decrease in the overall cardiac arrests rates, deaths after cardiac arrest, hospital mortality and failure to rescue" (Chen et al., 2016).

What we see in this Between the Flags program is the need for ongoing vigilance coupled with the readiness to intervene whenever clinical signs move outside the levels required to maintain homeostasis. When the signs are just outside the white, safe zone—in the yellow zones—prompt review and assessment are required, perhaps leading to a specific intervention or correction, and perhaps not. When the signs move into the red zones, immediate action is required: there is no longer any doubt that the clinical problem is acute and potentially serious. These three levels correspond exactly to what we have been suggesting about ethical awareness and ethical action, and about the need to intervene when clinical actions lead to responses outside the normal and morally expected

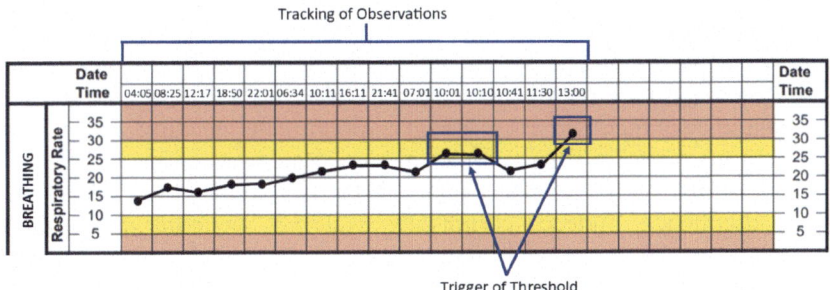

Fig. 8.1 A section of a clinical chart for the vital sign of breathing rate. The chart is color-coded into the three zones (white, yellow, and red) used in the Between the Flags program. As explained in the text, the white zone indicates stable observations, the yellow zones indicate observations that have shifted outside the homeostatic zone, and the red zones indicate observations that reflect dangerous levels. © NSW Clinical Excellence Commission 2020

course of events. To adopt the language that we used in discussing the Fletchers in Chapter 1, more-focused clinical attention and potential intervention are required whenever the level of clinical care, including both human/social/ethical factors and "technical" factors, is not "good enough" (Sidebotham, 2017).

- *White zone.* In the normal and morally expected course of events, the competent execution of a skill (very broadly construed, as we've indicated before, as the *implicitly ethical action* of providing care to a patient, communicating with the family, or interacting with colleagues) requires no special attention other than to continue along the same, broadly defined clinical course. While doing so, however, the clinician continues to monitor for potential discrepancies relative to expectations—which is itself an exercise of ethical awareness.
- *Yellow zone.* A minor discrepancy is detected, requiring some adjustment, in which context the touchstone questions (Text Box 8.1) will come into play. The discrepancy and requisite adjustment may be straightforward (e.g., changing a medication or engaging in a more extended conversation with a patient who has initially refused some obviously needed treatment) or may require more careful thought (e.g., consulting with a senior colleague about an unusual clinical development or deciding when and how to talk to a more junior nurse about keeping the medical staff informed about changes in a patient's condition).
- *Red zone.* A significant discrepancy is detected—the sign of serious failure, conflict, or even danger, potentially needing immediate action to protect the patient, family, or staff from harm. Immediate attention to the touchstone questions is required, and it's of no consequence—indeed, completely irrelevant—whether the underlying problems might ultimately prove to be ethical, "technical," or a combination of both. Due attention to the touchstone questions, if necessary in consultation with senior colleagues or experts of one kind or another, should lead to the resolution of the clinical situation. Note that one resolution of the clinical situation, as in the case of a terminally ill patient, may be to let the clinical situation continue to evolve, potentially leading to the patient's death or to their transfer to a palliative care setting (see next section).

What we hope this discussion of the Between the Flags program helps to communicate is that the ethical dimensions of clinical health care are always present as part of the action framework (see Chapter 6) that serves to shape the delivery of good, competent clinical care. That is, these ethical dimensions frame our actions, shape our understanding of the normal and morally expected course of events, determine how we monitor clinical situations for potential discrepancies, and come into play as we analyze and address discrepancies and the underlying problems via the touchstone questions. And throughout all this, the ongoing ethical component, rather than being in the foreground, may remain embedded in the clinical situation, in our perception of discrepancies, and in how we think, feel, and act. When things are going well, the ethical component will remain in the background. When something goes wrong, the ethical component will potentially come to the fore.

Engaging the Whole Person

An Example from Palliative Care

We see the above dynamic at work in a recent, 2024 article in *BMC Palliative Care*, which sets out just what happens when clinicians encounter a discrepancy in relation to the normal and morally expected course of events.

The goal of the study was to "explore how palliative care staff (including doctors, nurses, and social workers) reason about the autonomy challenges that arise when a patient who has first said he wants full information appears to change his mind and rejects being informed" (Björk, 2024, p. 1). To relate this situation to our discussion above, the normal and morally expected course of events was defined by the patient's preference for full information and by the expectation that palliative care staff would therefore provide the patient with all relevant information about his condition. But then the patient changes course—a departure from what was, for that patient at that time, the normal and morally expected course of events—and specifically states a preference for *not* being informed.

Though the hypothetical situation is clear, the palliative care staff's responses to it proved to be much different and much more varied from what the researchers were expecting. The hypothetical situation presented to the six different teams of caregivers was explicitly designed to present an "ethical dilemma" concerning autonomy, which the researchers expected would generate something like a standard set of responses built around autonomy and more or less falling within standard bioethical discussions of such matters. Instead, however, the participants and teams

- made no effort to "make a structured analysis" of that dilemma,
- rarely used the words *ethics* and *autonomy*, and
- did not conceptualize the situation as involving a conflict between autonomy and beneficence. (Björk, 2024, p. 8)

Even more importantly for our purposes, the participants and teams looked at the situations in their broader institutional and personal contexts, just as we would expect based on the approach presented in this book. That is, they considered all the following:

- Whether the patient actually wanted to reject any further information (despite the patient's clear, competent statement to exactly that effect)
- The possibility that the patient's preferences could change again (and therefore should regularly and gently probed over time in the context of the evolving conversation and relationship) and should not be relied on as necessarily invariant
- Their own "striving" to provide optimal clinical care and to achieve what they perceived as the patient's "good death"
- Their own interests, such as seeing themselves as having "provided the help they felt was necessary"
- The interests and rights of others, such as the patient's children (Björk, 2024, pp. 5–6)

Finally, the participants consistently looked to their own feelings and frustrations when assessing what they should do, and why. In effect, they were relying on the touchstone process, coupled with their own working moral frameworks, in assessing the hypothetical situation and their appropriate responses to it. That is, rather than looking at the situation as necessitating "top-down" assessments requiring the application of ethical principles such as autonomy and beneficence, the participants took a "bottom-up" approach in which their own feelings and preferences set the stage for determining how to respond, and what they could potentially hope to achieve, in the situation presented.

The Socio-Institutional Milieu

As the *BMC Palliative Care* article makes clear, questions of autonomy (for example) don't arise in a vacuum, with all questions organized and answered along a single dimension:

> When faced with a choice to withhold information as per a patient's wishes, or to provide information with the patient's best interest in mind, staff find it hard to balance competing values. Staff also find it hard to balance their own interests against a purely professional stance. The overall strategy seems to be to look for caring ways to impart the information. (Björk, 2024, p. 1)

The socio-institutional context thus comes dramatically into play here. Among other things, we see the participants taking into account (in addition to the patient's express articulation of his rights) the patient's best interests (as perceived by the clinical staff), the staff's perception of a "good death" (as understood within the confines of a palliative care service), the staff's perception of optimal care for this patient (as defined by the institutional goals of the palliative care service), the staff's perception of what they themselves wanted to achieve as individual clinicians (an expression of their own values and own work ethic), the rights and interests of others, including children (a reflection of the patient's own social milieu), and, more broadly, the staff's own ongoing feelings and reactions to the clinical situation, including the patient's medical condition and preferences.

What these various responses to, and perceptions of, the hypothetical clinical situation reveal is that the participants approach their work, and make judgments, as whole persons, with thought, feeling, and action fully engaged. Included in this context are their understandings of their roles and responsibilities; the dimensions of their moral experience that have been developing lifelong; and the action frameworks that they bring to everything they see, feel, and do. It is consequently as whole persons, and within specific socio-institutional milieus, that clinicians engage in informal ethical discourse and explore the ethical dimensions of their work.

CLINICAL ENGAGEMENT AS MORAL ENGAGEMENT

In our view, the foundation for all clinical ethics is the care and treatment of the patient: what should be done in the first place and what should be done, in turn, when that care and treatment somehow lead to unexpected results or fall short? As we've seen above, this way of understanding clinical ethics involves no distinction between nominally technical tasks and skills, on the one hand, and the human, interpersonal, and institutional dimensions of patient care, on the other. The central questions are the same: In the care of any individual patient, what kind of care is appropriate, and what should the outcome look like? If something discrepant occurs, what went wrong, and why? Within the bounds of the possible, how can the situation be fixed, and how can it be avoided in the future?

Similar questions may arise in the context of working with colleagues. Handovers, team meetings, consultations with other teams or clinicians from other fields (e.g., a consultation between medicine and nursing), and supervisions of various sorts may all go well (as expected) or poorly (not as expected), and potentially require adjustments of various kinds. Just which adjustments are needed will depend upon what has gone wrong and why.

More broadly, this same mode of analysis can be used to understand group processes within clinical settings. Case conferences, including Morbidity & Mortality Rounds, are specifically organized to determine, in the case of adverse events, what happened and why, and what needs to be done to improve patient care. All the questions we asked above concerning discrepancies between intentions/goals and outcomes in the work of individual clinicians can be asked in group settings and with a more systemic orientation.

Consider the following vignette, from the second author (KK), involving a case conference organized at her hospital (adapted from Scher & Kozlowska, 2020):

My hospital holds periodic, case-oriented ethics conferences, which are run by the Ethics Officer. In organizing one of these conferences, and in framing the questions to be addressed, the Ethics Officer completely avoided ethical language or terminology of any kind. The discussion at the conference itself turned out to be lively and focused, and—importantly—not a single ethical term was used, and not a single ethical principle was mentioned, during the entire conference. The discussion simply focused on what had happened, why, and what needed to be done in the future. At the end of the conference, the bioethicist from the local university was asked to comment, as she always does in one form or another, to provide some learned, informed ethical analysis. When asked this time, however, she said that she had nothing to add because all the issues had been very well covered. In my opinion this particular ethics conference was the best that had ever been held during my many years at the hospital.

What we see in this example is exactly what we would expect, given our analysis of informal ethical discourse, the working moral framework (in tandem with the action framework), and the need to address discrepancies between expectations and outcomes. Within the working moral and action frameworks of health care clinicians, and within the broader settings of clinical health care, discrepancies are perceived and experienced simply as clinical problems. And these problems need to be addressed using the touchstone process (and, in particular, the touchstone questions) in relation to the operative goals of clinicians and the institutions in which they work.

Looking more broadly at the challenges of health care education and training, consistently asking students and trainees to attend to the touchstone process (Fig. 7.1)—is a straightforward, intuitive, and learnable way of enabling them to address problems in clinical ethics. The touchstones for learning and touchstone questions—if asked—require no special vocabulary or training, either for teaching or learning. They can be seamlessly integrated into clinical teaching, training, and practice. And as trainees and clinicians respond to the touchstone questions (Text Box 8.1/7.2), bring into play the information thereby generated, and seek a way forward, they rely on what we have called the foundational assets that forge the clinician's voice. In doing so, they also take into account the socio-institutional factors that we have discussed above and that frame both the problem and the available solutions.

In the following chapter we look at these same issues in the context of prospective action: clinicians' decisions to change the course of care; program design; application of policy; and conflict resolution.

REFERENCES

Andersen, L. W., Holmberg, M. J., Berg, K. M., Donnino, M. W., & Granfeldt, A. (2019). In-hospital cardiac arrest: A review. *JAMA, 321*(12), 1200–1210. https://doi.org/10.1001/jama.2019.1696

Björk, J. (2024). "It is very hard to just accept this"—A qualitative study of palliative care teams' ethical reasoning when patients do not want information. *BMC Palliative Care, 23*, Article 91. https://doi.org/10.1186/s12904-024-01412-8

Chen, J., Ou, L., Flabouris, A., Hillman, K., Bellomo, R., & Parr, M. (2016). Impact of a standardized rapid response system on outcomes in a large healthcare jurisdiction. *Resuscitation, 107*, 47–56. https://doi.org/10.1016/j.res uscitation.2016.07.240

Fish, S. E. (1980). *Is there a text in this class? The authority of interpretive communities*. Harvard University Press.

Fish, S. E. (1989). *Doing what comes naturally: Change, rhetoric, and the practice of theory in literary and legal studies*. Duke University Press.

Hampshire, S. (1983). *Morality and conflict*. B. Blackwell.

Hughes, C., Pain, C., Braithwaite, J., & Hillman, K. (2014). 'Between the flags': Implementing a rapid response system at scale. *BMJ Quality & Safety, 23*(9), 714–717. https://doi.org/10.1136/bmjqs-2014-002845

Kleinman, A. (1995). *Writing at the margin: Discourse between anthropology and medicine*. University of California Press.

Kleinman, A. (1999). Experience and its moral modes: Culture, human conditions, and disorder [lecture delivered in 1998]. In G. B. Peterson (Ed.), *The Tanner lectures on human values* (Vol. 20, pp. 355–420). University of Utah Press; Cambridge University Press. https://tannerlectures.org/lectures/exp erience-and-its-moral-modes-culture-human-conditions-and-disorder/

Rawls, J. (1971). *A theory of justice*. Belknap Press of Harvard University Press.

Scher, S., & Kozlowska, K. (2018). *Rethinking health care ethics*. Palgrave Macmillan.

Scher, S., & Kozlowska, K. (2020). Teaching ethics in psychiatry: Time to reset. *Harvard Review of Psychiatry, 28*(5), 328–333. https://doi.org/10.1097/HRP.0000000000000258

Sidebotham, C. (2017). Good enough is good enough. *British Journal of General Practice, 67*(660), Article 311. https://doi.org/10.3399/bjgp17X691409

CHAPTER 9

Prospective Action and the Language of the Clinic

Abstract We have just looked at the role of informal ethical discourse and the touchstone questions in responding to discrepant experiences (Chapter 8). In this chapter we look at the role of informal ethical discourse and a parallel set of concrete clinical questions in determining the course of *prospective* action. In such situations we need to identify our intentions, existing clinical goals, the risks and benefits of the available options, and potential paths for going forward. We look at how this process works out regarding treatment plans, program design, application of policy, and conflict resolution.

Keywords Action framework · Application of policy · Clinical ethics · Conflict resolution · Embedded ethics · Health care ethics · Informal ethical discourse · Program design · Prospective action · Touchstone questions · Touchstones for learning · Safe and Sound · Working moral framework

In the previous chapter we discussed how the touchstone questions are used by clinicians to understand and respond to problems that have arisen in delivering health care to patients or in communicating with families or colleagues. We also discussed the Between the Flags clinical tool, which is used to mark "in the here and now" whether a patient's vital signs fall within or outside the normal and morally expected course of events, and to indicate whether immediate corrective action is required (Hughes et al., 2014). In this chapter we focus on forward-looking situations and prospective actions: clinicians' decisions to change the course of care; program design, including the response to recurrent or systemic problems in patient care; application of policy; and conflict resolution.

The same sorts of concrete clinical questions (i.e., the touchstone questions) that we asked in the previous chapter regarding *past actions* and their potentially discrepant outcomes can and should be asked *prospectively* in the context of assessing the available options and deciding on a particular course of *future* action. Among other things, these questions need to identify the range of available actions and assess the potential risks and benefits of those options, including the existing constraints, whether individual, social, or institutional. See Text Box 9.1 for a list of potential questions to ask in such circumstances.

Text Box 9.1 Questions Regarding Prospective Action

- What are my/our goals, or what do I/we aim to achieve? Are these goals or aims consistent with one another?
- What are the relevant goals of the patients or families involved? Are these goals or aims consistent with one another?
- Do my/our goals, or my/our means of attempting to achieve them, potentially conflict with the goals or interests of others (e.g., patients, families, colleagues, or administrators)?
- If so, how do I/we take to resolve such conflicts or take those other persons into account?
- Are there systems/institutional factors that I/we need to take into account?
- What are the potential means of achieving what I/we want—that is, what are the alternative courses of action?
- How do the uncertainties or risks associated with certain courses of action affect how I/we balance the advantages and disadvantages of the available options?
- To assess and balance the available options, do I/we need additional information?
- What are the potential impediments to achieving what I/we want to achieve?

– Can I/we make the decision unilaterally, or do I/we need to negotiate or otherwise collaborate or work with other concerned actors?

– If the decision is mine alone, how do I reach a balance among the potential factors and potential risks or uncertainties involved?

– If the decision is not mine alone, how do we reach an agreement or consensus as to how to move forward, especially since different actors may not want to achieve exactly what I do or may reach a different balance of factors, risks, and uncertainties?

– What are the relevant professional and institutional standards and rules?

– What are the best clinical practices, including as defined in recent systematic reviews?

– Do I/we need to undertake any consultations, if available, with senior colleagues or clinical or outside experts?

– In the case of recurrent or systemic problems, do I/we need to trigger a broader institutional effort to understand and address the problems?

PROSPECTIVE ACTION

Changing the Course of Care

In Chapter 6 we saw that the action framework, taken to include the working moral framework, has a dual role—first as shaping the background expectations against which incoming experience is assessed as expected or unexpected, and second as shaping how we respond to that experience, including any unexpected outcome. The action framework likewise has a dual role regarding *prospective* action. In the case of *ongoing* action, we simply stick with the action if it is unfolding as expected (analogous to the situation when a skill is performed successfully, as mentioned in Chapter 8 and to be discussed in detail in Chapter 10), even if minor adjustments are required. But the situation is different if our past action is now complete or if a past action has not proceeded as expected, raising questions about whether that past (and potentially ongoing) course of action needs to be reconsidered and potentially changed. It is at this point that the questions presented in Text Box 9.1 come into play. Of special note, the existence of a recurrent or systemic problem might be considered as the basis for a broader, potentially systemic review of the clinical situation.

In balancing various factors one against another and in reaching a determination concerning any particular course of action, the clinician inescapably relies on the action framework itself (Chapter 6) and on the dimensions of moral experience that shape each clinician's thinking, feeling, and doing (Chapter 5). For further discussion of this process, see Chapter 12.

Program Design

At The Children's Hospital at Westmead—and Sydney Children's Hospital Network, more broadly—an ongoing project (headed by Anne Preisz, the Clinical Ethics Manager) has been drawing on the expertise and experience of clinicians throughout the hospital and network to design a program for managing recurrent conflicts between health professionals and families. The project is called "'Safe and Sound': A Graded Response to Managing Challenging Behaviour Between Health Care Professionals (HCPs) and Parents/Carers" (Basu & Preisz, 2023; Preisz, 2020).[1] Its goal is to identify parents' progressively challenging/ aggressive behaviors and to set out a series of "unified strategies" that will enable clinicians to "continue supporting families while setting clear, safe, cohesive boundaries" (Preisz, 2020). The overarching aims are to protect the staff and to guide their behavior while also promoting the interests of patients and families. As is typical of ethical problems within clinical health care, the ongoing challenge in articulating *Safe and Sound* program guidelines has been to flesh out a careful, thoughtful balance between the many potential factors in play.

The *Safe and Sound* project has developed two matrices (visual representations) for clinicians to use in guiding prospective action. Each of the matrices sets out three levels of family behavior, ranging from low-level, anticipated conflict (as occurs in the presence of "Heightened emotions, confusion and feeling a loss of control") to outright *aggression*, with the middle level of *challenging* behaviors falling between the two.

To see how this works, we'll use the first of the two matrices as our example. It presents the recommended staff responses (broadly described) to the family for each level of family behavior. After giving examples of the behaviors characteristic of each level, the matrix sets out the particular actions to be taken and also suggests language to be used in dealing with the family. To illustrate how the matrix works, we'll look in detail at the middle level, challenging behaviors (Preisz, 2020):

- example family behaviors: "Micromanagement/demanding/refusing care, Indirect verbal abuse/insults/swearing, Passive aggression/intimidation" that "Upsets or threatens other families/other patients, Impedes care of the patient"
- general characterization of staff response: "Support and define clear boundaries"
- specific actions (partial list): "Acknowledge and validate the emotional response; Step away from the perceived threat to physically delineate a boundary; Follow up and close the loop: Outline the inappropriate behavior and consequences; Seek agreement on behaviour, in writing if necessary; Collaborate with [multidisciplinary team] to ensure unity"
- suggested language (partial list): "We both need to work together to best care for your child; I can see/hear you are upset/angry; I am feeling uncomfortable because . . . ; What you are doing/saying is not OK; You need to step out of the unit and take a break"

The ultimate impact of this matrix (and also the second) is to turn difficult situations, each of which could easily spin out of control, into predictable scenarios with suggested responses. The graded responses outlined in the matrices amount to a new framework for advancing each clinician's repertoire of behaviors for addressing conflicts with families.

In the context of our discussion of informal ethical discourse and its being embedded in the values and goals of clinical health care, what is especially noteworthy here is that the two matrices reflect the considered judgments of the hospital's clinical staff. With the assistance of the *Safe and Sound* team, the clinicians have been able to identify concrete clinical problems, establish realistic goals, and determine the most likely means of achieving them. In the process, the clinicians have been addressing such matters as

- their own levels of comfort with, and tolerance for, conflict,

- their perceptions regarding what could reasonably be expected of families experiencing potentially extreme stress,
- their determinations as to the baseline interests of patients and how to protect and promote those interests under difficult circumstances, and
- their understanding of what actions are open to clinicians in each particular clinical setting.

In discussing all these matters, the clinicians have needed to continually balance their own goals and values, both within their own thinking and in relation to others, all with the goal of reaching some sort of working consensus. This entire project has been, in effect, a group exercise in posing and seeking answers to wide-ranging clinical problems via the questions listed in Text Box 9.1—an example of informal ethics and clinical reasoning for prospective action at their best and most probing.

Application of Policy

We can see the same processes at work in the following vignette involving a transplant committee's process for selecting recipients from among the candidates presented by treating teams:

My first year as a renal fellow was incredibly exciting. In addition to seeing renal physiology up close, I was involved in six renal transplants—a record for a first-year trainee! The renal meetings, where we discussed the pros and cons of transplantation for each patient, were especially engaging. I came to understand that deciding which patients would be considered eligible for transplants at our medical center involved a complex web of interrelated factors. Tissue type and likelihood of success were obvious factors, but then we considered matters of age, overall health, urgency, and family (including support) circumstances, among many potential other factors. In presenting each patient to the transplant committee, which made the actual allocation decisions, we included all relevant factors and how we had assessed them.

Near the end of my fellowship, I presented the team's work at an academic conference. During the question period I was asked what ethical principles our team applied in discussing transplant candidates. My answer was that generally, we didn't. Our routine was to discuss in depth and at length each patient's situation from a biopsychosocial perspective. We tried our best to tease out the biological/medical, psychological, and social issues and then to put them together (not always without some initial conflict) in a balanced way that captured a working consensus. Later on, after the conference had finished, I reflected that the team's biopsychosocial approach allowed us to navigate through complex situations and address difficult issues, including ethical ones, in our own familiar language of clinical medicine.

The reader will no doubt notice that in daily clinical practice, this discussion and comparison of candidates for kidney transplants—a problem that has received an astonishing amount of attention in the bioethics literature—consistently uses the language of *informal* ethical discourse, which is often further embedded (see next main section) in the compressed language of clinical health care.

Conflict Resolution

Conflict—both between persons and within the thinking of any particular person or group—might well be considered one of the "signature" phenomena of modern health care ethics. And it's commonly thought that it's only through the explicit use of formal ethical discourse—the language of bioethicists, not clinicians—that one can properly understand, analyze, and address such conflicts. Our emphasis throughout this book, however, has been on informal ethics and on the social and institutional embeddedness of clinical ethics. We further elaborated the notion of informal ethics via the touchstone questions regarding past/ongoing action introduced in Chapter 7 (Text Box 7.2) and the parallel clinical questions regarding prospective action introduced in this chapter (Text Box 9.1).

In efforts to understand conflict of any sort and to determine how to resolve it, our experience is that clinicians rely on versions of these two sets of questions, with their concrete clinical and factual focus. What they don't rely on is the formal ethical discourse that they may have learned about during their earlier professional education and training but that subsequently falls out of use as unconnected with the clinical milieu or with how clinicians actually think and act within it. The questions presented in Text Boxes 7.2 and 9.1 embody the language of the clinic and its unrelenting need to address disputes as they arise—and in terms that clinicians, either as individuals or in groups, can readily understand and that they have long experience in working with. Notably, these questions, with their embedded elements of informal ethical discourse, do not distinguish between "technical" and "ethical" problems. All clinical problems can be, and are, equally addressed via the clinical questions presented in Text Boxes 7.2 and 9.1.

In the following vignette (an amalgam), we see how clinicians resolve conflict by working with and through the goals and constraints of the immediate clinical situation, which requires in this case a clear understanding of the multiple factors that potentially compromise the course of care.

Donald was 12-year-old boy presenting with functional neurological disorder (FND): functional (stress) seizures occurring many times a day and loss of power and coordination in his legs (requiring a wheelchair). Donald's parents were long separated and mutually intolerant of each other. They were also distrustful of others and generally reluctant to put the care of their child into the hands of medical professionals, especially when, most recently, a nurse in the Emergency Department had insinuated that the problem was all in Donald's head.

During the assessment by the mind-body team, the atmosphere was tense, and the parents initially resisted providing a developmental history (an essential part of the team's assessment process). The father was also adamant that no medications were to be used at any point for any purpose. The team psychiatrist—the most senior team member—acknowledged this request and noted that the Emergency Department nurse had spoken out of line. She also explained that Donald's medical situation reflected a neurophysiological response to ongoing stresses, as described by the parents during the assessment itself.

The two-week inpatient intervention under the Mind-Body Program started badly. The psychiatry resident working with the mind-body team inadvertently mentioned to Donald that at some point he might benefit from some form of PRN medication. When the father heard about this from Donald, he exploded. This outright disregard of the father's wishes played directly into his distrust of others (and especially those in authority), and it confirmed, for him, that the mind-body intervention was not for his son. This attitude was further enforced by the father's recent internet searches, where he had discovered numerous websites with personal accounts depicting FND as inescapably chronic, with the consequence that little good could be expected of the present treatment.

At a scheduled meeting on the fourth day of the mind-body intervention, the father brought up all these grievances, and noted that the money he had previously spent on Donald's psychotherapy had apparently gone for naught.

It was clear to the team that the father was challenging the team and reenacting the history of endless conflict between the parents. If the team was unable to address the grievances, he would likely withdraw his consent, jeopardizing Donald's treatment. There was no time to plan a response; we needed to respond immediately, in the moment.

The team psychiatrist asked the father whether her resident had apologized for his mistake. While acknowledging that the resident had apologized, the father reiterated that it was not OK and that it never should have happened. The psychiatrist agreed but also noted that they all had a common interest in moving forward and helping Donald. If the resident offered a second apology, would that be enough for the father? This new apology was offered and accepted, allowing the team and father to focus again on Donald and his progress.

The psychiatrist then validated the father's sense of frustration: how hard it was for a parent to watch a previously healthy child struggle with FND symptoms and to manage the uncertainty as to what would happen next. She noted—in a clear, matter-of-fact way—that the internet accounts that the father had watched pertained to adults and that the accounts did not apply to children. The majority of children attained complete recovery. And complete recovery was, they all agreed, the goal of the current treatment intervention with Donald. She also noted that the work that Donald had done with his psychologist had not actually gone to waste. Donald had engaged in the program with ease, was working hard, and making daily progress. The entire team recognized that Donald's previous psychotherapy had made this all possible.

In response to the above, the father's anxiety settled, as did his anger and frustration with the team. He was now eager to support Donald and all his work in completing the treatment program. On discharge, Donald's legs were working normally; he was discharged with a functional seizure safety plan; and his functional seizures were increasingly short and less frequent.

Unexpectedly, Donald's father contacted the team a couple of weeks after Donald's discharge. The discharge plan was working well, and Donald was happy to be back at school.

As is apparent from this vignette, the technical and ethical problems are interwoven, and the solutions emerge via pragmatic thinking about the clinical situation, the relevant actors (patient, family, and clinicians), and the likely most effective way of surmounting the problems and moving ahead clinically. The team psychiatrist was also forthright in acknowledging that mistakes had been made, thereby crediting the father's emotional response and allowing them all to leave the mistakes behind.

The Compressed Language of Clinical Health Care

As we have seen above and through this book, the ethics of health care, whether medicine or nursing or social work or clinical psychology, is embedded in the setting—broadly conceived to include professional training, the institutional setting in which care is provided, the relationship between clinicians and patients, and the ongoing need to collaborate with other health professionals—and, indeed, in the clinicians themselves. As befits a professional environment in which time constraints are real and everywhere, and in which decisions often need to be made in a hurry and on the spot, the language of clinical health care and its embedded ethics is often highly compressed.

A tangible example of this compressed language, repeated thousands of times a day in some form or other, can be seen in what happens when a doctor, we'll call her Dr. Smith, enters an examination room, with the patient waiting:

'Hello. I'm Doctor Smith. What's troubling you?

If this example were filled out in formal ethical discourse (see Chapter 2)—of course, no one would ever speak like this—it would look something like the following:

Hello. I'm Doctor Smith. As a doctor, I am obligated to provide the best medical care to my patients that is available in the immediate practice setting. You are now my patient, and we are here. I therefore have an obligation to provide you the best medical care available in this setting. In order to do that, you need to start by telling me what's troubling you.

Rather than providing a formal analysis of the doctor's ethical stance toward the patient, the compressed, informal query by Dr. Smith captures the essence of the doctor-patient relationship and its embedded ethics. What Dr. Smith has done is, in effect, to promise to her patient that she will bring her professional experience and skills to bear on whatever problem the patient presents. This "simple" promise brings into play, however, an implicit promise for Dr. Smith to use all her substantive knowledge and technical skills acquired through medical school and thereafter, her social and communications skills for dealing with patients and her colleagues, her knowledge of the institutional setting (including both its capacities and its limits), and her understanding of the care available under current financial arrangements. These implicit commitments are all contained in the simple utterance "What's troubling you?".

This process of engaging the patient is, for any health professional, largely implicit, and it may verge on the nonverbal. The patient in the above example may say, "It's my arm," or may simply point at an obviously injured limb or finger. A nurse may obtain patient cooperation in some matter or other by saying a single word, giving a gentle nudge, or simply raising her eyebrows. A clinical psychologist, pushed hard to do something that would violate professional boundaries, might simply say, "I can't" (with or without an explanation), just shake his head, or simply scowl.

Compressed encounters such as these—with patients and also with colleagues and staff—occur dozens of times a day in the life of any health care clinician. The amount of talking will vary widely from patient to patient and with the complexity of the problem presented. What remains consistent, however, is that (1) these encounters all have a baseline ethical component (involving, e.g., the clinician-patient relationship, the goals of health care and how to achieve them, or the need for clear, effective communication), and (2) next to no explicitly ethical language will be used in addressing the clinical situation, including any potential problems or conflicts, even when both the clinician and the patient recognize the ethical aims and constraints of the clinical encounter and are knowingly trying to act in accordance with them. The ethics of health care is embedded in ordinary, non-ethical language such as "we need to" or "we can't" or "the patient has decided" or "the risk to the patient is too high." The ethics remains embedded in the situation and its discourse unless and until something goes wrong or something unexpected happens to throw off the normal give-and-take between clinician and patient or clinician and colleague. Even then, the language used is that of informal ethical discourse. See Text Box 9.2 for typical linguistic expressions from informal ethical discourse and their equivalents in formal ethical discourse. See also the clinical questions listed in Text Boxes 7.2 and 9.1.

Text Box 9.2 The Language of Clinical Health Care

Typical clinical language (informal ethical discourse)	*Equivalent language from formal ethical discourse*
The patient needs *x*	The patient (or client) has an ethical right to receive *x* OR We have an ethical obligation to provide *x* to the patient (or client).
What does the patient want? Does she have the information she needs?	We have an ethical obligation to respect the patient's autonomy. We are therefore ethically obligated to provide the patient with all relevant information, discuss that information with her, and determine what she has decided to do.
Doing it *that* way will hurt the patient. It's better to do it *this* way	The principle of nonmaleficence ethically mandates that we avoid doing harm. Doing it *that* way will cause harm to the patient. We therefore have an ethical obligation to do it some *other* way. Since *this* other way will serve the same purpose but cause less or no harm, we will do it *this* way.
I think *x* would help the patient	Since *x* would help the patient, the principle of beneficence ethically requires us to do *x*.
We need more information	It would violate our ethical obligation of due care to the patient to decide on a course of action without obtaining more information.
We need to consult (either with a specialist or a supervisor)	None of us currently involved in this patient's care is sufficiently expert to determine how to proceed in this difficult case. We therefore, as a matter of promoting the best interests of the patient, as required by the ethical principle of beneficence, are ethically obligated to call in a specialist consultant.

(continued)

(continued)

Typical clinical language (informal ethical discourse)	Equivalent language from formal ethical discourse
We've been doing x in all these cases	Since we have previously done x in all similar cases, the principle of justice ethically requires us to do x in this case in the absence of any factor that distinguishes this case from the previous ones.
We need to talk to the patient and potentially sit down with the family, too	We are ethically obligated to tell the truth to our patients and also to families (subject to the patient's consent). The patient and family have made every effort to avoid confronting the truth about the patient's condition. We have an ethical obligation to correct that situation.
She doesn't want them to know	We are ethically obligated to keep patient information confidential. She has specifically stated that she does not want her family to be informed about her condition. We are therefore ethically prohibited from providing that information to her family.

Finally, note that this compressed language is characteristic of any clinician's own thinking, separate from any exchanges with others. When something is discrepant, the internal dialogue might be as simple as "Am I going to do that? No." And at the other extreme, the internal dialogue might mimic an extended discussion with others, involving progressive questioning, answers, and consideration of consequences—all the elements of informal ethical discourse discussed above and embedded in the clinical setting.

IMPLICATIONS

Clinical problems necessarily arise within what we have called clinicians' action frameworks and, more broadly, within their own frameworks of thought, feeling, and action, as summarized at the outset of this chapter and presented in the preceding two chapters. These are inherently pragmatic, goal-directed frameworks, and that's exactly how problems in clinical health care—whether ethical or non-ethical—are identified, analyzed, and solved.

Clinical ethics is inescapably embedded in every clinical interaction and in every clinical process. Good clinical care *is* ethical clinical care. As we noted above, clinical problems are just that: clinical problems. When something discrepant occurs, the challenge is simply to find out what happened and why, and to correct the situation in any way that's appropriate. And when a clinical situation, unrelated to discrepant outcomes, presents numerous possible courses of action, the challenge is to identify the best way of moving forward, taking into account the full range of clinical factors in play. Addressing these challenges inescapably involves the use of concrete clinical questions such as those presented in Text Boxes 7.2 and 9.1. This form of conceptualization and analysis plays directly into the strengths and commitments of clinicians, and also into their daily, ongoing efforts to provide good clinical care. The focus is on the clinical situation—what needs to happen to provide good clinical care.

The implications for clinical problem-solving, for teaching and learning clinical ethics, and for consulting in clinical ethics are unequivocal and straightforward.

Clinical Problem-Solving

As discussed above, clinical problems can range from those that are simple and straightforward, requiring only minor adjustments in behavior or clinical care, to the extreme, requiring careful thought and aggressive interventions. Many problems will fall somewhere in between, requiring some thought, potentially some discussion, and careful adjustments in behavior or clinical care. In all these situations, the focus is typically on the immediate demands of the clinical setting. Whether dealing with patients or colleagues, and whether dealing with past, present, or prospective actions, the clinical questions from Text Boxes 7.2 and 9.1 point the way forward, both for understanding the present situation and for determining what needs to be done to address it.

Formal ethical discourse (see Chapter 2) may be useful in some situations, though it will be useful only if it connects up directly with the experience and perception of the clinicians involved; the usefulness of formal ethical discourse for clinical problem-solving needs to be demonstrated in any particular situation rather than simply assumed. Our view is that formal ethical discourse should be brought in (e.g., via an ethical consult) only if clinicians are unable to solve the problem on their own, including through consultations with colleagues, group discussions, or case conferences, and that every effort should be made to connect up such formal ethical discourse with the informal ethical discourse of clinicians.

Teaching and Learning Clinical Ethics

The needs of health care trainees need to be determined prospectively by taking into account the skills that they will need to solve clinical problems in the years ahead. As we've seen here, the capacities to identify, analyze, and solve clinical problems, including those in clinical ethics, are tied into the capacity to use the touchstone process (Fig. 7.1) and questions from Text Boxes 7.2 and 9.1 to examine clinical situations within the action framework. The aim is to determine (1) what went wrong, why, and how to repair the situation, or (2), looking prospectively, what needs to be done in order to achieve the operative clinical goals. Engaging in this reflective process requires teaching and scaffolding if health care trainees are to develop and use the requisite skills. When this scaffolding is not provided—when trainees' foundational assets are not nurtured during the training process—trainees' baseline human responses and their caregiving skills potentially erode (see Chapter 12 in Hojat [2016] for a review). Since the quality of patient care ultimately depends on clinicians' bringing into play the full scope of their human capacities, the demonstration and teaching of these reflective skills in daily clinical practice is of crucial importance for long-term professional growth.

Given that informal ethical discourse is already embedded in the concepts and goals of clinical health care (and in the lives of clinicians themselves), the teaching and learning of clinical ethics need to utilize informal ethical discourse and should not rely on formal ethical discourse, which often does not further advance clinical discussions or clinical understandings. Formal ethical discourse translates informal ethical discourse into an abstracted form that, for clinicians tied into the concrete world of health care and its incessant demands, is often both difficult to understand and burdensome to work with.

Consulting in Clinical Ethics

Consultants in clinical ethics would likely make a more effective, generative, and lasting contribution to local health care settings, to the work of clinicians, and to the welfare of patients—if that is their aim—by using informal ethical discourse to engage directly with health care's local moral processes. Included in this context is the need to engage with clinicians' foundational personal assets, as discussed in Part II of this book: their knowledge of roles and responsibilities, their working moral frameworks, and their action frameworks. Problems need to be addressed in the language of the clinic and by engaging the thoughts, feelings, and actions of the clinicians working there (see, e.g., the *Safe and Sound* project above). As with clinical problem-solving, the relevance of formal ethical discourse needs to be demonstrated rather than assumed.

A final, crucially important point concerns the different ways that clinical problems or situations can be perceived, understood, and addressed. For clinicians, problems arise as discrepancies (regarding past actions) or indeterminacies (regarding prospective action) within their action frameworks. And they arise, more broadly, within the overarching frameworks of thought, feeling, and action that clinicians bring to each of their clinical encounters and to their ongoing interactions with patients, families, and colleagues. As such, the contours of each clinical situation are defined by those concrete, real-world details about both the clinical situation and each particular clinician. Likewise, these concrete, real-world details will determine how each problem or situation is perceived or experienced, discussed, and solved (as by generating questions focused on what, why, and how). As we saw with the Between the Flags program (Chapter 8) and the Safe and Sound project discussed above, ongoing discussions with clinicians were the means of identifying and analyzing clinical problems and of determining, in the ever-so-concrete world of the clinic, the best way forward, all things considered. To remove the discussion of such problems from their immediate interpersonal and clinical contexts is to change the problems themselves. Such broader discussions, abstracted from the clinical context—such as those that philosophers or bioethicists may pursue—may or may not be interesting or useful to clinicians. But they must be understood as falling somewhat wide of the mark in addressing the immediate clinical needs of the clinicians involved.

NOTE

1. The final version of *'Safe and Sound': A Graded Response to Managing Challenging Behaviour Between Health Care Professionals (HCPs) and Parents/Carers* has yet to be published. The quoted material is drawn from a 2020 poster by Anne Preisz, the team leader, as made available on New South Wales Clinical Excellence Commission website, at https://cec.health.nsw.gov.au/__data/assets/pdf_file/0007/609280/Cohort-23-ECLP-Poster-Anne-Preisz.PDF (last accessed 23 March 2024). The PDF is no longer available on that website, but readers are encouraged to write Anne Preisz at SCHN-ClinicalEthicsSupport@health.nsw.gov.au for the latest version of the matrix.

REFERENCES

Basu, S., & Preisz, A. (2023). Family conflict and aggression in the paediatric intensive care unit: Responding to challenges in practice. *Clinical Ethics,* *18*(4), 410–417. https://doi.org/10.1177/14777509221091081

Hojat, M. (2016). *Empathy in health professions education and patient care.* Springer International Publishing.

Hughes, C., Pain, C., Braithwaite, J., & Hillman, K. (2014). 'Between the flags': Implementing a rapid response system at scale. *BMJ Quality & Safety, 23*(9), 714–717. https://doi.org/10.1136/bmjqs-2014-002845

Preisz, A. (2020). Safe and sound. A graded response to managing challenging behaviour between Health Care Professionals (HCPs) and parents/carers. Executive Clinical Leadership Program (ECLP). Poster, Cohort 23. https://cec.health.nsw.gov.au/__data/assets/pdf_file/0007/609280/Cohort-23-ECLP-Poster-Anne-Preisz.PDF (accessed 23 March 2024; now removed from website).

CHAPTER 10

Expectations and Discrepancies

Abstract In this chapter we look at how clinical skills develop over time, at the diversity of skills to be learned, and at the inherently social and interactive character of the learning process. Learning to identify when something has "gone wrong"—that is, to identify discrepancies from what was expected—is a crucial element of this process of learning clinical skills and continuing to improve those skills over time. We also explain why and how ethics pervades the performance of all clinical tasks, as well as why attention to the touchstones for learning and touchstone questions serves as a means of understanding and addressing the ethical challenges and problems of the clinical milieu.

Keywords Clinical competence · Clinical ethics · Discrepancies · Dreyfus model of skill acquisition · Embedded ethics · Expectations · Expertise · Health care ethics · Learning process · Normal and morally expected course of events · Novice · Professional competence · Professional development · Skill acquisition

In the three preceding chapters on moral engagement in clinical health care, we showed how predictive coding, sensitivity to feeling states (including homeostatic [gut] feelings), and our understanding of the normal and morally expected course of events enable us to identify discrepancies in our experience. We then saw how these processes lead us to pose the touchstone questions (Text Box 7.2) and questions regarding prospective action (Text Box 9.1) that enable us to understand and address the problems, including the ethical problems, that arise in clinical health care. We saw, too, that the language for addressing these problems is the often-compressed language of informal ethical discourse.

In this chapter we build on those results by looking at the relevant processes from a distinctly different perspective. Rather than looking at how problems arise and are understood in the here and now—at any particular cross-section in time—we look at these processes as they develop and change over time in the life of any trainee or clinician. We consider how the process of engaging in clinical tasks, time and time again—from the very outset of each trainee's clinical experience—generates expectations, enables the identification of discrepancies, and *sets the stage* for addressing clinical problems, including potentially ethical problems, right there and then. What results, over time—if these skills are nurtured and developed—is a highly nuanced, ever-growing capacity to use the touchstone process as a tool for learning and to identify, understand, and address problems in clinical health care. That is, the process of becoming an "insider," as described in Chapter 3, runs in parallel with the progressive growth, step by step, of competence in managing clinical tasks and scenarios, interpersonal challenges of all sorts (involving colleagues, other health professionals, patients, and families), and potentially questions of clinical ethics.

The Advanced Beginner

In Chapter 3 we discussed the growth of clinical competence in terms of the Accreditation Council for Graduate Medical Education's Milestones project. Here we use what has come to be known as the *Dreyfus model* of skill acquisition as a way of thinking about the same process, specifically in relation to nursing. That model, with its five stages of competence— novice, advanced beginner, competent, proficient, and expert—was originally articulated by Stuart and Hubert Dreyfus in a 1980 report prepared for the U.S. Air Force Office of Scientific Research. The model was then expanded in a 1986 book, *Mind over Machine: The Power of Human Intuition and Expertise in the Era of the Computer.* Patricia Benner—a prominent nursing theorist and educator, and now professor emerita at the University of California, San Francisco School of Nursing—used this model in *From Novice to Expert: Excellence and Power in Clinical Nursing*

Practice (1984), to frame her research on the professional development of nurses.

Of special interest for us is the stage of advanced beginner, which is characteristic of senior nursing students or newly graduated nurses. In a recent article, Benner (2022) presents the following conversation between two nurses at the advanced beginner stage:[1]

Nurse 1: I think what's stressful is the expectation on us to be good all the time, and have all that knowledge right now.

Nurse 2: Right. Because it is not just one system that is failing. And these patients get real critical when everything is going on, and you have to think about everything that could possibly go wrong. And everything is going wrong. And it's like the knowledge should be there, and where is it? And it's not. And if it is there, it is in the notes, but you still have to read it, and you don't have time right now to get out that piece of paper say: 'Okay, dig[oxin] toxicity. This is what I should look for, and this is what I should do.'

This exchange captures the situation of any trainee at the outset of their clinical training in any field of health care. At that stage, trainees typically possess a potentially substantial core of "textbook knowledge" and a smidgen of experiential or clinical knowledge. What is striking, however, is the incapacity to form reliable, determinate predictions about the future—a product of not having mastered the procedures or health care challenges in question, but also a product of not knowing what might go wrong and why, and of not knowing, in turn, what interventions would be required to address whatever problems or complications might arise. What is happening? Why? What should I do? How much? How long? What's next? What am I missing? What have I forgotten? What else do I need to know? What else might happen? Who is going to help me?

So begins the long, gradual process of coming to understand and manage the normal and morally expected course of events. Benner notes, as suggested above, that the first steps are tied in with identifying particular procedures or interventions that need to be completed. But such a description underplays the diversity and complexity of what needs to be accomplished, and how.

The Diversity of Skills to Be Learned

The exchange above mentions the effort to deal with digoxin toxicity, but the range of tasks or situations that a beginning nurse (or beginning doctor; see Chapter 3) needs to master is staggering. The Elsevier Clinical Teaching (for nursing) website lists 166 "procedures," each of which has many steps or dimensions (Elsevier, 2024). And as Nurse 2 notes above—in relation to working with a modicum of clinical experience and knowledge—"you have to think about everything that could possibly go wrong. And everything is going wrong" (Benner, 2022). Likewise, the New South Wales (Australia) intensive care unit (ICU) nursing clinical pathway divides the pathway into Dreyfus/Benner stages and presents the skills that need to be mastered at each stage (Intensive Care NSW, 2022).

In this New South Wales nursing clinical pathway, one of the skills included at the stage of Advanced Beginner Level 1 is "Underwater seal drainage management and troubleshooting," also known as "chest drain management." As is apparent from Text Box 10.1, this deceptively simple description is shorthand for a wide range of skills and actions (Royal Children's Hospital at Melbourne, 2022).[2]

Text Box 10.1 Royal Children's Hospital at Melbourne, Nursing Guidelines for Chest Drain Management (Main Headings and Subheadings Only)

Introduction
Aim
Definition of Terms
Indications for Insertion of a Chest Drain
Insertion of a Chest Drain
Chest Drain Assessment & Management
Start of Shift Checks
UWSD Labelling
Patient Assessment
  Vital signs
  Forward areas
  Drain insertion site and dressing assessment
  Skin integrity
Pain UWSD unit and tubing

Suction
Drainage
Volume
Colour and Consistency
Air Leak (bubbling)
Oscillation (swing)
Other Considerations
Patient Positioning
Patient Transport
Specimen Collection
Chest Drain Dressings
Dressing Change
Anchoring Drain Tubing
Removal of Dressings
Changing the Chamber
Removal of Chest Drains
Patient and Pre-procedure Preparation
Post Procedure Care
Complications and Troubleshooting
Pneumothorax
Bleeding at the Drain Site
Accidental Disconnection of System
Accidental Drain Removal
Purse String Cut or Not Present
Unable to Remove Chest Drain
Retained Drain During Removal
Family Centred Care

Clinical Learning as Inherently Social and Interactive

In examining these various clinical guidelines, goals, and requirements, the inherently social and interactive character of the learning process jumps out at you. Virtually everything that is learned needs to be learned via communication with peers, senior colleagues, patients, families, and other health professionals, coupled with observations of all of these. A nurse does not, for example, learn to troubleshoot chest drains in isolation. It takes clinical judgment—the product of clinical observation, plus the interpretation and guidance provided by more senior nurses and one's other colleagues—even to identify what constitutes a problem, and then one needs to learn from others what needs to be done to address each problem. Interactions with patients and families are ongoing and, especially in an ICU setting, potentially difficult and stressful.

In general, in order to learn any particular task, the trainee needs to do some or all of the following:

- listen carefully to, or observe, the person teaching the task in question
- clarify any potential questions
- perform the task itself (making, in process, any minor adjustments that might be needed)
- determine whether the task was completed successfully
- interpret the result
- discuss the result with the teacher/supervisor
- determine what should be done next

Further, if the task is one involving a procedure, the trainee needs to do some or all of the following:

- explain the procedure to the patient (and family members)
- answer any questions
- obtain informed consent
- report the result of the procedure, with an interpretation (if appropriate) to the patient (and family members)
- discuss what should be done next

The care and deliberateness involved in mastering any particular procedure is apparent in the following vignette, told by a patient, concerning the insertion of an in-dwelling, Foley catheter.

At a rural hospital, I needed to have an in-dwelling catheter removed and a new one reinserted. The nurse assigned to do the procedure was what is called in Australia an "enrolled nurse." Though he was nearing the point of completing his separate qualifications as a registered nurse—a much higher level of competence—he had obviously not had extensive (or any?)

experience inserting Foley catheters. When he was ready to proceed (after setting up everything on his cart), he rolled the cart to my bedside, with every piece of equipment meticulously laid out. His level of self-conscious deliberation was palpable. At each step, I could see him looking over the cart and identifying just what he needed at that point. Carefully picking it up, he would then explain to me exactly what he was doing, and he would then do it—slowly and deliberately. After each step, he would pause, take another look at the cart, and then proceed again, as before. And so it went—for about, by my recollection, roughly 20 minutes, possibly even more. I was especially aware of the pace and time because, a week earlier at the same hospital, a much more experienced nurse had inserted a catheter smoothly and seamlessly in what could have been no more than a few minutes. It was going to be some time, and many repetitions, before the enrolled nurse achieved the same level of competence.

What is ever so clear in this vignette is that the process of learning procedures and skills involves slow, painstaking, ongoing practice and dedication, with the goal of bringing those skills to the level of professional competence, embedded in procedural memory.

Clinical Challenges and Cumulative Learning

Another series of challenges comes into play if a task has *not* been performed successfully. In that case, the trainee needs to do some or all of the following:

- discuss the situation with the teacher/supervisor
- determine what went wrong, and why
- determine what needs to be done differently
- negotiate a repeat attempt with the patient (if a patient is directly involved)
- plus, do much of what was required for the first attempt, as described above

And still further, to achieve competence in any particular skill or task, the trainee needs to learn to identify and address predictable complications and other foreseeable problems. In the case of chest drain management, for example, these situations include pneumothorax, bleeding at the drain site, accidental disconnection of system, accidental drain removal, and inability to remove the chest drain, among other problems (see Text

Box 10.1). The trainee also needs to learn how to work and communicate effectively, in the context of such complications and problems, with patients, families, and other health professionals.

Finally, note that all the above elements deal with the nurse's immediate, first-order responses that then generate additional responses from others—the patient, family, and other health professionals. The nurse then has to determine, in turn, how to respond appropriately to those responses. And so on. The vignette below illustrates the second-order issues that can arise in the task of inserting a nasogastric tube with a patient in an eating-disorder program.

I have worked as a bedside nurse all of my nursing career. The skill of inserting a nasogastric (NG) tube was taught in my first year of nursing— the "fundamentals of nursing." The first step of learning involved the use of a mannequin. When using the mannequin we mouthed each step of the procedure out loud. We focused our attention on what we were doing and practiced in a slow, mindful, and methodical way. Our intention was to get the procedure right so that when we performed it with a patient, we would do it correctly, thereby avoiding any problems or pitfalls. After the mannequin, the next step was supervision by a clinical facilitator with a consenting adult patient. Five attempts were allowed. I remember my first patient. I focused my concentration on the task at hand, bringing each step carefully to mind, and telling the patient what I was doing—and what they needed to do (when to swallow)—moment by moment. At this point in the learning process, it took me at least 20 minutes to set up and to complete the procedure. With time, I mastered the technical elements of the procedure, and I could insert an NG tube in less than 5 minutes.

Some years later, when I began to work with young people with eating disorders, the insertion of NG tubes took on new layers of complexity, and this time, I was on my own, with no text to consult and no supervisor to guide me. Even when they had given consent, young people with eating disorders were highly ambivalent about the insertion of the NG tube. They were slow to swallow; they adopted body postures that hindered insertion; and they sometimes began to sneeze, cough, spit, or thrash their heads after the procedure had been started. In the eating disorders setting, a successful NG insertion therefore required an additional set of skills: negotiating with the patient to obtain effective cooperation; offering sedation to help reduce distress; using sensory strategies to help divert focus-of-attention; establishing a clinical team (usually 2 to 4 people) to support the process; and at times, using a traffic-light system that enables the patient to communicate (by pointing) her or his level of rising arousal or distress.

In thinking about the series of actions, responses, and counter-responses discussed above (as in the case of learning chest drain management or how to insert a nasogastric tube), what needs to be kept in mind is that (1) mastering a single skill or set of skills is a multifaceted task, (2) this multifaceted task concerns only one of hundreds of multifaceted tasks/skills that trainees and clinicians are expected to master, and (3) each of these skills presents opportunities and challenges regarding the problem-solving, social, and communication skills required to achieve and maintain competence.

A key question in this context is whether the training process preserves what we see as the human element of the skills being taught. As we've argued, students, trainees, and clinicians who have been taught to use the touchstones for learning and, more generally, the touchstone process (Fig. 7.1) are inescapably looking at their clinical experience at the interface of affect and intellect, and by the same token they are in a position to understand and address, if properly encouraged and supported, the human dimensions of caring for patients. When the touchstones for learning are ignored, or when the human dimensions of patient care are otherwise downplayed or dismissed, the educational process yields students, trainees, and clinicians who are well-trained technicians but who have limited skills in providing empathic and holistic clinical care, where communication and the skill of "caring" play a central role (Hojat, 2016; Kleinman, 2019; Novack, 1995; Pellegrino, 1979).

VARIATIONS: EXPECTED AND UNEXPECTED

This process of building skills generates, for any particular skill, a set of expectations—in effect, a normal and morally expected course of events for that skill. When the execution of that skill leads to consequences that fall *within* the range of what was expected, then the clinician proceeds with whatever needs to be done next. But if the consequences are *outside* the range of what was expected, then the touchstone questions come into play. The clinician needs to take notice, determine what happened and why, potentially consult with peers or supervisors, discuss the situation with the patient, and do whatever else is required to move the process ahead. But this initially *unexpected* consequence then comes to be integrated, with time—or at least we should hope—into what is, for that particular clinician, their own normal and morally expected course of events. When so integrated to reflect previous learning, the now-expected (previously unexpected) occurrence is addressed, often more or less automatically, with any minor adjustments being made on the fly.

For each clinician, for each skill, this process is ongoing, and lifelong. Early on—as with Nurses 1 and 2 earlier in this chapter—scarcely anything is normal and morally expected. They simply didn't know enough to make sound projections; hence, their anxiety. But with increasing experience and knowledge, spread over many skills and many challenges of one kind or another, the two nurses—like any other young clinicians—will come to experience what was initially experienced as surprising or even shocking, as falling within a set of expectations. "Oh, my god!" gives way to "OK. So now we need to . . ." and a clear path of action. We can see this dynamic at work in an incident from John Berger's *A Fortunate Man* (1997), a book in which Berger bears witness to the work and life of John Sassall, a country doctor working for Britain's National Health Service.

> After receiving a call that a woodsman was trapped beneath a tree, his leg mangled, Sassall rushed to the scene. "As he drove through the lanes he kept his thumb on the horn the whole time, partly to warn oncoming traffic, partly so that the man under the tree might hear it and know that the doctor was coming" (p. 17). Once Sassall was at the injured man's side, the men "looked at the doctor. His hands are at home on a body. Even these new wounds which had not existed twenty minutes before were familiar to him." (pp. 17–18)

It's not that Sassall was expecting anything in particular—other than a seriously injured leg. And though all such injuries differ one from another, they generally fall within the range of normal and morally expected, requiring only minor adjustments from one case to the next. And so it is with all the skills, whether technical or social/interpersonal, that health care professionals acquire over the course of their careers.

Seeing the familiar in the unfamiliar, as in the above vignette, is a capacity that grows with experience, as we see in the following vignette. Note that the term *old digger* is Australian slang for soldiers who served in World War I.

> Decades ago, when I was working as a surgical resident at a veterans' hospital, I was called to the bedside of an old digger who had stabbed himself. I found him lying down, comfortable, talking, with a bread-and-butter knife sticking out of his chest to the left of the sternum,

waving in time with his heartbeat. The consultant who treated him was a thoracic surgeon of great experience but not a formally trained cardio-vascular surgeon. In theatre we opened the chest to find the knife well embedded in the wall of the ventricle, waving with each systole but with no bleeding around it. My consultant put a purse string suture around the site, and I pulled the knife out as he tied the suture. I learned that the heart, though an organ with special characteristics, is ultimately just another organ. If there's a hole in it, sew it up.

Many years later, as the on-call surgeon at another hospital, I was called to the Emergency Department to see a boy of about nine who had been stabbed in the chest by his paranoid schizophrenic mother in the belief that they were both about to die violently. Copious blood was coming from his intercostal catheter, and the boy was at a critical level of circulatory collapse. We went straight to theatre. A right thoracotomy (entering the side of the chest cavity via a cut between the ribs), revealed the chest half full of blood. Gently feeling around the chest cavity, my finger located a hole in the right atrium of the heart. When I told the theatre staff, they became very agitated, and reacted by demanding we call a cardiac surgeon. But there was no time. By feel, I applied a long artery forceps to each end of the hole, which allowed the application of a curved vascular clamp to stop the bleeding. It was then a simple matter of sewing the hole closed.

Early in his training, this young surgeon saw how the senior surgeon's mature clinical judgment enabled him to identify a simple solution to what was, for the young surgeon, an unfamiliar and potentially fraught clinical situation. Years later, he exercised the same sort of mature clinical judgment in coming to realize that the patient's life-threatening clinical situation presented the same problem as the case from many years before: the need to close a hole. But this deceptively simple solution was possible only because of the surgeon's long clinical experience and armamentarium of skills developed through years of training and clinical work: his recognition of urgency, his knowledge of anatomy, the skill in his hands to feel out the problem, his technical skills and knowledge of surgical instruments, and his emotional capacity to stay calm and to keep the operating team calm and focused.

Notably, as experience increases in relation to clinical skills, so does the trainee's or clinician's familiarity with the expected responses of patients, families, peers, and other clinicians in any particular clinical situation. That is, just as one's mastery of clinical skills increases over time, so does one's capacity to interact effectively with others in an effort to achieve the desired clinical goal. Embedded in the clinical milieu, trainees and young clinicians become increasingly familiar with what is expected of them and what can be expected of others. What was new and surprising at first will become part of the normal and morally expected course of events. These newly recognized elements of the normal and morally expected course of events might range, of course, from the emergent, at one extreme— requiring an immediate, focused response (as in the vignettes involving John Sassall and the surgeon above)—to the entirely routine, at the other (as in determining whether a wound can be closed with a Steri-Strip rather than a stitch).

At the center of this increased predictability is the broad mastery of the fundamental skills defining professional competence in any field of health care. Deciding whether one has executed a skill correctly, however, potentially requires some clinical judgment, and the process of fine-tuning any particular skill might be ongoing over one's entire clinical career; see Text Box 10.2 for a useful example (Pharmaguideline, 2021). In any event, a determination of "correct execution" provides a crucial, and extremely helpful, bit of information regarding any potential problem or deficiency, ethical or otherwise. If a procedure or process (such as a clinical interview) has been performed properly, but the results of that procedure or process do not fall within the normal and morally expected course of events, then the touchstone questions come immediately into play. That is, the determination of correct execution serves as a reliable basis for looking around—at other circumstances, other persons, and other potential failures, lapses, or complicating factors. Combined with one's cumulative learning regarding patients, families, peers, senior colleagues, and other clinicians, along with increasing knowledge of the clinical setting itself and how it works, the clinician is in a position to start generating hypotheses about exactly what happened, why, and what needs to be done to correct the problem (technical or interpersonal).

Text Box 10.2 Fine-Tuning Basic Skills

It is not always easy to determine, early in one's clinical career, whether a skill has been executed correctly or not. Checking information in a database may be straightforward, as may many low-order technical procedures in nursing or other fields. The basic problem is that, in addition to a standard procedure that is taught to all relevant trainees and considered adequate—"Yes, you've done it correctly!"—many procedures also are understood as having "fine points," or minor improvements that one is likely to learn only with further reading or experience or through specific advice or supervision from someone more experienced. Just how extensive these "fine points" can be is apparent in the case of administering medications.

Learning to fine-tune the dosing of any particular medication for any particular patient is not necessarily straightforward. For example, the website Pharmaceutical Guidelines lists 18 separate factors that potentially influence the dosing of medications (© Pharmaguideline 2021. Reproduced with permission):

- Age
- Sex
- Body weight
- Time of administration
- Body surfaces area
- Route of administration
- Emotional factors
- Accumulation
- Environmental factors
- Presence of disease
- Additive effect
- Idiosyncrasy
- Tachyphylaxis
- Antagonism
- Tolerance
- Metabolic disturbances
- Drug dependence/addiction

No early trainee (e.g., an advanced beginner who has made significant progress toward competence) will know to make adjustments for all the above factors or even know that the above elements are relevant to dosing decisions. But if a question arises about the level of dosing and whether it may potentially have some impact on the drug's effect, a senior clinician may eyeball the situation and suggest that one or another of the above factors may need to be considered. A senior clinician may be able to take such factors into account by simply looking at a patient and determining, even without conscious thought, whether any of the above factors may be relevant to the medication regimen for the patient in question. But the naturalness or ease with which senior clinicians make such judgments belies a long course of learning about these factors and how to adjust for them, often one by one and often through the direct supervision of mentors.

What goes for dosing goes, to some degree, for any other skill. There's the basic skill, and there's the additional information and skill required to fine-tune it, thereby eliminating potential complications or minor errors and effectively further individualizing the care or treatment.

We can see exactly this phenomenon—and the value of correct execution—in the following vignette.

Our team assessed a 10-year-old girl referred by the general pediatric team for admission into our mind-body rehabilitation program for the treatment of functional seizures. On assessment the child's events were characterized by increased respiratory rate (hyperventilation), whole-body shaking, a wide-eyed fear expression, and brief periods of "zoning out." The events had been fully investigated and, at the time of presentation, looked like functional seizures. Some days into the admission, the episodes had failed to show any therapeutic response to treatment as usual, and they changed in character. The child began to experience confusion, altered levels of consciousness, and motor activation with extreme levels of arousal, all lasting over many hours. These features were not typical of functional seizures. Noting the discrepancies, our mind-body team decided that the initial diagnosis was mistaken, so we sought help from neurology. Following repeat medical investigations—the same as had been done before—the working diagnosis of functional seizures was changed to antibody-negative inflammatory encephalitis comorbid with a neuropsychiatric syndrome. The joint neurology-psychiatry team and the child's parents then embarked on a long and arduous course of care (see Hawkes and colleagues [2024] for a detailed description).

To reiterate the point of this vignette, the mind-body team knew that they had conducted their assessment and initiated their treatment with all due care, and they also had the clinical experience to know how functional symptoms, including functional seizures, commonly present and how they respond to treatment. Against this background, they were able to determine that the child's presentation was likely not functional and therefore that she needed further neurology assessment. They were also able to hypothesize that one element or another of the standard neurological workup may have returned a false-negative finding. In sum, given that they knew that their own work had been competently performed, the mind-body team were in a position to ask what may have gone wrong and what needed to be done.

THE PERVASIVE ETHICS OF CLINICAL CARE

All this talk about competence masks the deeper reality that this process of ongoing professional growth—acquiring clinical skills, dealing with patients, families, peers, and colleagues, and identifying and addressing problems of all kinds that arise in the clinic—is inescapably ethical. For example, if one looks at the nursing guidelines for chest drain management (Text Box 10.1), what it sets out is a list of skills, but what it also sets out, in effect, is the set of ethical obligations for nurses who are responsible for chest drain management. Exactly the same could be said of any list of skills and subskills defining a particular task in nursing or in any other field of health care. Any effort to perform any of these skills is inherently ethical, as are any efforts to understand or address any clinical problems or deficiencies that may arise. Clinical ethics is inescapably embedded in every clinical interaction and in every clinical process. And in thinking about these interactions and processes, clinicians use the concrete language of the clinic, making use of the touchstones for learning and touchstone questions, and always keeping the immediate clinical goals in mind (including in relation in prospective action).

As noted in Chapters 7 and 8, the touchstones for learning are tied in with affective experiences in response to discrepancy—experiences that lead to questions that engage clinicians' informal ethical discourse, duly expressed in the language of clinical health care. It is the concrete, contextualized language of the clinic that enables clinicians to focus best and most directly on what has gone wrong, why, and what needs to be done to address the situation (and to avoid such problems in the future). Although it would be possible to introduce the language of formal ethical discourse into such discussions, our own view is that doing so diverts clinicians' attention away from what they know, concretely and in detail, and presents them with the more abstract, effortful task of thinking and solving problems via the unfamiliar idiom of academic ethics. The overall effect would be counterproductive.

We return here to our analysis of informal ethical discourse embedded in Fish's interpretive communities (Fish, 1980, 1989), Hampshire's ways of life (Hampshire, 1983), and Kleinman's moral modes of experience (Kleinman, 1995, 1999) (see Chapter 2). Each of these conceptions of social/professional settings defines a concrete world in which clinicians work and act, think and feel. As clinicians learn to understand these worlds and what is, in each, the normal and morally expected course of events, they learn responses that incorporate the informal ethics embedded in the setting, in their interpretive community, and then in themselves. They also—via the range of pragmatic questions legitimated by the setting and its goals—develop further mastery of the setting and its ongoing demands, including its embedded ethical demands and goals.

NOTES

1. This exchange, posted on Nursingdevelopment.com, a professional development website created by Benner, is a minor revision of an exchange that originally appeared in the second edition of Benner's *Expertise in Nursing Practice: Caring, Clinical Judgment, and Ethics* (2009, p. 32).
2. The text box summarizes the main headings and subheadings from this website. For content detail, please access the website itself (Royal Children's Hospital, 2022).

REFERENCES

Benner, P. E. (1984). *From novice to expert: Excellence and power in clinical nursing practice*. Addison-Wesley Pub. Co., Nursing Division.

Benner, P. E. (2009). *Expertise in nursing practice: Caring, clinical judgment & ethics* (2nd ed.). Springer International Publishing.

Benner, P. E. (2022). Teaching and learning implications of the Dreyfus and Dreyfus skill acquisition model. *Educating Nurses*. https://www.educat ingnurses.com/from-novice-to-mastery-ii-the-dreyfus-and-dreyfus-model-of-skill-acquisition-in-nursing-practice

Berger, J. (1997). *A fortunate man: The story of a country doctor*. Vintage International.

Dreyfus, H. L. (1986). *Mind over machine: The power of human intuition and expertise in the era of the computer*. Free Press.

Dreyfus, S. E., & Dreyfus, H. L. (1980). *A five-stage model of mental activities involved in directed skill acquisition.* https://apps.dtic.mil/sti/citations/ADA 084551

Elsevier. (2024). *Clinical Skills.* https://www.elsevierclinicalskills.co.uk

Fish, S. E. (1980). *Is there a text in this class? The authority of interpretive communities.* Harvard University Press.

Fish, S. E. (1989). *Doing what comes naturally: Change, rhetoric, and the practice of theory in literary and legal studies.* Duke University Press.

Hampshire, S. (1983). *Morality and conflict.* B. Blackwell.

Hawkes, C., Dale, R. C., Scher, S., Cornish, J. L., Perez, D. L., Santoro, J. D., Fernandes, S., Kozlowska, K. (2024). Bridging the divide: An integrated neurobio-psycho-social approach to treating antibody negative inflammatory encephalitis in a school-aged child. *Harvard Review of Psychiatry, 32*(3), 101–116. https://doi.org/10.1097/HRP.0000000000000395

Hojat, M. (2016). *Empathy in health professions education and patient care.* Springer International Publishing.

Intensive Care NSW. (2022). *Nursing clinical pathway.* https://aci.health.nsw. gov.au/networks/icnsw/clinicians/nursing-clinical-pathway

Kleinman, A. (1995). *Writing at the margin: Discourse between anthropology and medicine.* University of California Press.

Kleinman, A. (1999). Experience and its moral modes: Culture, human conditions, and disorder [lecture delivered in 1998]. In G. B. Peterson (Ed.), *The Tanner lectures on human* values (Vol. 20, pp. 355–420). University of Utah Press; Cambridge University Press. https://tannerlectures.org/lectures/exp erience-and-its-moral-modes-culture-human-conditions-and-disorder/

Kleinman, A. (2019). *The soul of care: The moral education of a husband and a doctor.* Viking.

Novack, D. H. (1995). Therapeutic aspects of the clinical encounter. In M. Lipkin, S. M. Putnam, A. Lazare, J. G. Carroll, & R. M. Frankel (Eds.), *The medical interview: Clinical care, education, and research* (pp. 32–49). Springer.

Pellegrino, E. D. (1979). *Humanism and the physician.* University of Tennessee Press.

Pharmaguideline. (2021). *Definition and factors affecting posology.* https://www. pharmaguideline.com/2021/08/definition-and-factors-affecting-posology. html

Royal Children's Hospital at Melbourne (2022; updated October 2022). *Chest drain insertion guideline.* https://www.rch.org.au/rchcpg/hospital_clinical_ guideline_index/chest_drain_management/#FCC

CHAPTER 11

Two Modes of Clinical Ethics

Abstract As an alternative to the "standard mode" of clinical ethics, with its typical emphasis on formal ethical reasoning (often including ethical principles) and other forms of "top-down" analysis, we suggest that clinical ethics is better understood in terms of socio-institutional processes. It is just such processes that are at work in Stanley Fish's interpretive communities, that engage students as they begin their professional education, and that frame how trainees and clinicians feel and think about the ethical challenges and problems that they encounter in their work. We also look at how these socio-institutional processes are interconnected with the systemic, human, and social elements of institutional culture in professional schools and hospitals.

Keywords Clinical ethics · Cycle of institutional change · Ethical principles · Ethical reasoning and analysis · Formal ethical discourse · Health care ethics · Informal ethical discourse · Institutional culture · Socio-institutional ethics · Socio-institutional processes

In the Introduction to this book, we mentioned something that a psychiatrist friend of ours, Danielle Vandenberg (1961–2019), said after reading our first ethics book, *Rethinking Health Care Ethics* (Scher & Kozlowska, 2018):

© The Author(s) 2025
S. Scher and K. Kozlowska, *Revitalizing Health Care Ethics*,
https://doi.org/10.1007/978-3-031-78475-0_11

> I can't tell you how many time my colleagues and I were discussing some problem of medical ethics and ended the discussion with "We need an ethicist." But now I realize that the ethicist is us!

This observation encapsulates, albeit somewhat cryptically, the central idea of this chapter and, indeed, the book: the ethical empowerment of clinicians. The core of the observation is that Danielle had previously come to see bioethicists as having the expertise to address difficult problems in health care, whereas she and her fellow clinicians, without the input of bioethicists, saw themselves as lacking the intellectual resources to reach determinate, well-reasoned solutions to such problems.

A BIFURCATED WORLD

What we see here is the *learned incapacity* of clinicians to address—in the absence of specifically ethical training or the explicit assistance of bioethicists—ethical problems encountered in clinical health care. This learned incapacity reflects the bioethicist's bifurcation of the clinical world into two domains. In one of these domains, the problems are fundamentally clinical and require a clinical intervention (that is, an intervention that requires the exercise of learned clinical skills). In the other of these domains, the problems are fundamentally ethical and require ethical analysis (that is, the use of ethical terminology and ethical principles or other concepts [such as obligations or rights]).

The following vignette articulates the bifurcated world as seen through the eyes of a physician and the physician's colleagues:

> A few years ago a new, publicly funded website was established to help health care professionals from all fields, and not just medicine, make better decisions involving clinical ethics. The launch of the website was widely publicized, and a renowned figure in bioethics was invited to present the inaugural lecture. I was one of a group of clinicians who, excited by this initiative in clinical ethics, organized a car share to get us to the lecture after finishing work. We sat together in the last row of the lecture hall. Our hopes of hearing something useful about the ethics embedded in our daily work began to ebb soon after the lecture began, when it became apparent that the lecturer was focusing exclusively on unusually complex, conflict-filled cases involving the treatment of terminally ill patients. The analysis was abstract and analytical, and unconnected from the way that

clinicians work with, or think about, most dying patients and their families. Even more off-putting was the lecturer's final suggestion that clinicians who managed patients at variance with his analysis were simply unethical. I noticed an immediate rise in my anger, and I thought of all the clinicians who worked day in and day out with patients who were dying: those who debrided their wounds; those who offered them food and drink; and those who answered them patiently when they (or their families) asked the same question for the nth time. I noted in my mind that the lecturer had constructed his entire analysis based on the tiny subset of cases where things go seriously awry. In this respect, I realized that he was following the bioethics tradition—with its problem-based focus-of-attention and exclusive interest in "ethical dilemmas" and how to analyze and solve them. Likewise, he omitted from his analysis the vast majority of cases that require an alternate perspective—a focus-of-attention that looks at the fundamental human dimensions of a situation and attempts to find a common ground for moving forward.

Over the course of the lecture, I could feel an emotional wave—a sort of collective grumble—flow through the audience. During the question period, some brave clinicians voiced their disagreement rather strongly, and all the questions from the many clinicians in attendance started out with some version of "things are actually a bit different clinically"—only to have the lecturer dismiss their questions and concerns as misguided. Though I was too much a coward to speak out, I wanted to ask how many dying patients the speaker had attended, how many wounds he had debrided, how many dying patients he had fed, how many crying families he had talked to. On leaving the lecture hall, one of my colleagues commented that the lecture was clinically of little value but—looking on the bright side—had at least "made us think." Another colleague noted that the occasion would have been very useful had the speaker not started his analysis "at the wrong end." I reflected that despite the organizers' obvious good intentions, the lecture starkly highlighted the discontinuity between the clinic and the academy. Academic ethics thrives on problems that are unusual and difficult. But clinical ethics is necessarily focused on the clinical situations that we encounter every day. For us practitioners, clinical ethics is about giving ethically appropriate care to the patient who is sitting in front of us.

Once one accepts the bifurcation between clinical and ethical problems, which is so apparent in the above vignette, the groundwork for ethically disempowering clinicians is effectively complete. The coin of the realm—at least in this particular, bifurcated realm—is taken to be ethical expertise or some near variant thereof, a top-down approach to teaching ethics and "doing" ethics (Kozlowska, 2022). The assumption is that bioethicists and ethics consultants possess some special capacity to make judgments about, or present recommendations concerning, challenging clinical situations, that clinicians lack this capacity, and that clinicians cannot acquire this capacity without enrolling in formal programs in bioethics or some equivalent program in bioethics consulting. What consequently happens is that professional schools in health care present lectures and sometimes also courses in health care ethics as part of the standard curriculum. The readings are typically drawn from the bioethics literature, with its own distinct vocabulary and approach to identifying, framing, analyzing, and solving problems in clinical ethics. An emphasis on the application of ethical principles continues as the predominant mode of conceptualizing health care ethics throughout the process of professional training and also in the bioethics literature itself. A further consequence is that training programs do not recognize the value (and long-term benefits) of promoting a bottom-up approach to clinical ethics (Kozlowska, 2022)—an approach that understands ethics to be embedded in every clinical situation (see, e.g., Chapters 8 and 9) and that respects and builds upon the informal ethical discourse that trainees inescapably bring to all their encounters with patients, families, and colleagues (see, e.g., Chapters 2 and 8). This failure to build on trainees' moral capacities has deep-reaching effects: the training process may decrease the trainee's capacity for moral engagement rather than increase it (Hojat, 2016; Kleinman, 1999).

We can see this emphasis on principles and abstract reasoning—a top-down approach to ethics—in the Milestones project of the United States' Accreditation Council for Graduate Medical Education (see Chapter 3). As mentioned in Chapter 4, for internal medicine the only mention of ethics in the entire set of milestones is in the section "Professionalism 2: Ethical Principles" (ACGME, 2020). That is, even at the level of graduate training in medicine, where the emphasis is on the nitty–gritty of clinical medicine, ethical principles are considered, in effect, the one and only means of understanding and addressing the ethical challenges in the practice of internal medicine. With this emphasis on ethical principles—even at the graduate level—it is no surprise that health care trainees and clinicians come to see ethical problems as requiring ethical analysis using formal ethical language. More broadly, the implication is that ethical problems can be analyzed and addressed only through the use of formal top-down ethical reasoning that overrides the pragmatics inherent in the clinical reasoning that clinicians use, and that they've mastered, in all other domains of clinical practice.

Some of the teaching in the top-down approach to ethics is, to be sure, fundamental to modern health care. Patient autonomy, informed consent, and respect for patients are essential, foundational elements of the clinician-patient relationship. Teaching students about such matters, along with the associated behaviors, is obviously a crucial part of professional education. But myriad other problems—such as those involving tensions and conflicts (e.g., in relation to patients, families, colleagues, administrators), inconsistent policies, stressful work environments, institutional problems and limitations, and the availability (or not) of particular treatments or other interventions to particular patients—are not. They are better addressed using the bottom-up approach described in this book. Rather than defining and exploring the foundational elements of *being* a clinician in any particular field of health care, these latter problems are ones that arise in relation to actually *exercising* or *working in* that role. Though professional *students* can talk in the abstract about day-to-day problems that confront *clinicians*, these problems (as they occur in clinical practice) are inherently fact-laden and involve particular people, ranging from the clinician to the patient, family, and any other clinicians or administrators involved, all with their particular personalities, goals, strengths, and weaknesses. Without the texture of a specific set of factual constraints, and without the immediate personal engagement inherent in the clinic but not the classroom, discussions of such matters are inescapably abstract and theoretical, disconnected factually and personally from the realities of clinical decision-making.

To take this analysis one step further, a weakness in efforts to determine the capacity of health professionals (and often, more specifically, doctors) to recognize ethical problems is that the setting for such research is not the clinic. A problem that has been constructed to test, in the abstract, a trainee's or clinician's capacity to recognize such problems—and by using the same language as that of the researcher—is simply not the same as a problem arising concretely, in the clinic and with one's own patient. A further weakness of such research is that, even in clinical practice, the existence of "an ethical problem" arising in relation to the care of any particular patient is of uncertain clinical utility. In the early 1980s, at a seminar run by the Hastings Center just north of New York City, a leading bioethicist noted to a senior physician practicing at a tertiary-care hospital in the City that he, the bioethicist, could identify an ethical problem in every single one of the physician's two dozen or so patients in the hospital. The doctor accepted the challenge and was then flabbergasted when the bioethicist did, indeed, identify an ethical problem relating to every single patient. The twist, though, is that despite now seeing that there was an ethical problem with every one of his inpatients, the doctor subsequently changed nothing in the care of any of those patients. That is, he remained satisfied with the care that each patient was then receiving, and each patient and each family continued to be satisfied with that care. The implication is that a bioethicist's conception of ethical problems in patient care does not necessarily map onto what clinicians would consider good, appropriate care or treatment.

Two Modes of Clinical Ethics

The "standard mode" of clinical ethics, the source of so much ire in the above vignette concerning the inaugural lecture for a clinical ethics website, is seriously at odds with the approach presented in this book. To be sure, and as we noted above, patient autonomy, informed consent, and respect for patients are foundational elements of any approach to health care ethics. Also foundational is an understanding of one's professional role as a member of any particular field of health care, such as medicine, nursing, clinical psychology, or social work. But what we want to focus on here is the difference between standard approaches to clinical ethics and the one presented here, or what we refer to as the *socio-institutional* approach.

In distinguishing between formal and informal ethical discourse in Chapter 2, we introduced three different ways of conceptualizing informal ethical discourse, which is, of course, the baseline natural language for all of us, including health care professionals and, indeed, bioethicists themselves. We mean for those three different ways—Fish's *interpretive communities* (Fish, 1980, 1989), Hampshire's *ways of life* (Hampshire, 1983), and Kleinman's *moral modes of experience* (Kleinman, 1995, 1999)—to be taken seriously and not just as providing an interesting background or context for the approach presented in this book. That is, we take these constructs as real and as defining the empirical baseline for ethical analysis. The ethics of clinical health care and the ethics of individual health professionals are interconnected with, and the product of interactions between,

- health care institutions (e.g., professional schools, hospitals, health care delivery organizations, and organized clinical practices),
- the embedded values and practices of each particular field of health care, and
- the ongoing social interactions (broadly conceived) between students, trainees, clinicians (of all sorts), patients, and families within health care institutions and in relation to those values and practices.

Clinical ethics, which is embedded in every clinical interaction and in every clinical process, needs to be understood in terms of this socio-institutional context.

Socio-Institutional Ethics

In Part II of this book—Forging the Clinician's Voice—we set out the foundations for the clinician's voice as it develops through professional education, training, and subsequent work as a mature clinician:

- lifelong familiarity with roles, responsibilities, and their organizational context, as presented in Chapter 4
- the ever-developing dimensions of moral experience, as outlined in the conceptual map presented in Chapter 5 and resulting in an evolving *working moral framework* for each clinician: fairness; autonomy and related concepts; truthfulness; trust/reliance/respect/reciprocity; doing good/avoiding harm; solidarity; and moral judgment
- the *action framework* presented in Chapter 6, incorporating each clinician's working moral framework and including salience and predictive coding, along with the capacities and skills that support the action framework: cognitive, perceptual, and affective capacities; communication skills; theory of mind; quest for narrative coherence; and capacity to recognize our limits

At the point of beginning professional school, students have developed, in effect, their own working moral frameworks and modes of informal ethical discourse (which vary from one person to another), as well as wide-ranging capacities for understanding their own intentions and goals and those of others.

The Socio-Institutional Encounter

A transformation occurs as health care students encounter the demands of their professional training. The forces that shape their ongoing professional development, including the development of their own working moral frameworks, action frameworks, and informal ethical discourse, are inherently social and institutional, and oriented toward the particular field of health care for which they are being trained. Referring to the eventual outcome of this process in terms of "clinical ethics" isn't technically incorrect. Nevertheless, this expression fails to capture the process itself, as well as the human outcome of this process. In *Doing What Comes Naturally* (see Chapter 2), Stanley Fish refers to "embedded practitioner[s] whose standards of judgment, canons of evidence, or normative measures are extensions" of the community itself (Fish, 1989, p. 144). Within health care, these distinctive processes are deeply interconnected with the professional training, professional standards (including the evidence base), characteristic interpersonal processes, and institutional settings of each particular field—hence our term *socio-institutional ethics.*

Given that students and trainees already possess working moral frameworks and action frameworks when they begin their professional studies in health care, what exactly is it that subsequently changes, enabling them to address the challenges of clinical health care? The setting in which they live and work changes in three ways: (1) Perhaps most obvious is that they are rapidly accumulating both factual and procedural knowledge of their own fields, whether, for example, medicine, nursing, social work, or clinical psychology. (2) The institutional setting moves progressively away from the classroom to the actual settings in which health care is delivered. (3) Likewise, the setting in which learning occurs moves progressively away from the classroom and library and toward settings that are inherently social and interactive—not just in working and interacting with colleagues on the ward or participating in rounds or case conferences of various sorts, but also in providing care to *patients.*

So are set in motion the processes discussed in the previous four chapters, including the following: the competent performance of skills; expectations and predictive coding; the recognition of discrepancies (via the touchstones for learning); and the need to ask the touchstone questions (retroactively, about past actions; Text Box 7.2) and the questions regarding prospective actions (Text Box 9.1). These processes—rather than some substantive change in ethical knowledge or reasoning as such—are the key to gaining competence in the socio-institutional ethics of clinical health care.

Institutional Culture: Systemic Elements

These processes do not take place in a vacuum. As noted, for example, in Chapter 10, training in clinical skills is intensely social and involves constant communication with peers, senior clinicians, patients, and families. And all this social interaction occurs within specific health care institutions, which have their own emphases, goals, constraints, and problems, all of which need to be taken into account, as necessary, when providing care to patients. A trainee follows—and there's always personal and institutional pressure to follow—the lead of more senior trainees and clinicians. And those clinicians, in turn, both set the tone for those who work under them and typically have assumed the tone—the professional emphases—of the institution itself, which is often why they've chosen to work at that particular institution in the first place. A teaching hospital is different, for example, from a hospital mainly oriented toward primary care, resulting in varying experiences for trainees. Each institution also has its own distinctive culture, with some priding themselves on their delivery of state-of-the-art health care and others priding themselves on their holistic care of patients and families (though some institutions endeavor to do both at once). And at the opposite end of the spectrum, when processes on multiple system levels have failed, lie institutions that have been unsuccessful in maintaining a viable culture and that have systematically failed to provide patients with good (including ethical) clinical care (Dyer, 2022; Vize, 2022).

The particular culture of each health care institution is usually simply embedded in the institution itself—following the work of Stanley Fish (see Chapter 2), in the very professionals who work there (Fish, 1980, 1989). But sometimes institutions endeavor to make that culture explicit. We can see such a process at work in the recent effort by Mass General

Brigham (MGB)[2] to develop *Operating Principles for How We Think and Work as an Integrated System of Care* (Mass General Brigham, 2024a). The Principles, or what we will refer to as Overarching (institutional) Goals, include the following: Trust and transparency; Patient-first design; System mindset; Collaborative behavior; Agile decision-making; Empathy and humility; and Personal accountability.[2] These goals are intended to guide people across all parts of the MGB system in working and collaborating as the system evolves—over time, in a rapidly changing health care environment—into a single, integrated system of care. Rather than being abstract and potentially difficult to apply, the Overarching Goals express tangible aims and constraints that serve as realistic signposts on the way to effective institutional decision-making. As MGB notes, they "are a tool for setting expectations within a working team and provide guidance for leaders and teams to model and share."

With a view to thus setting expectations and providing guidance, the Overarching Goals are supplemented by *Some Powerful Questions for Leading Our Transformation*—questions to be used to inform and frame decision-making that supports the goals (Mass General Brigham, 2024b). These questions address the assumptions being made, the potential need to consider new or alternative approaches, the possibility that broader system problems need to be included, the adequacy of the currently available information, the potential need to bring additional people into the discussion, the need to ensure that disagreement and dissent has been properly identified and addressed, and whether patients would see the background situation and potential solutions in the same way as the health professionals involved. These are the questions that administrators and clinicians need to ask as they consider specific problems or challenges that MGB encounters as a working institution. It is not a matter of providing care to any particular patient or determining how to deal with any particular clinician or other staff member. Instead, these questions concern organizational issues and matters of policy—how MGB functions as a system.

The reader will no doubt notice that the Powerful Questions bear an uncanny resemblance to what we introduced in Chapter 7 as the touchstone questions (Text Box 7.2) and the questions regarding prospective action (Text Box 9.1). This resemblance is exactly as it should be. What's at stake in both the Powerful Questions and our two sets of questions is the fundamental goal of advancing the interests of patients and aligning organizational and professional behavior and practices, including the work of all clinicians, in relation to that goal.

Institutional Culture: Human and Social Elements

What this characterization of the Overarching Goals and Powerful Questions leaves out is that it is only part of an ongoing cycle of institutional change involving institutional policies, institutional change, and the impact on, and role of, individuals living and working within the institution (See Fig. 11.1). At each stage, moreover, *individuals* within the institution play a central role. *Individuals* identify institutional problems and challenges. *Individuals* ask the Powerful Questions. *Individuals* are responsible for institutional operations, for implementing institutional policies, and for considering the Powerful Questions in view of the institution's Overarching Goals. And it is through the *individuals working together* within an institution that one comes to understand the impact of those policies and then, in turn, to understand whether those policies themselves present problems or challenges that need to be addressed. In every single step, the health professionals involved perceive the situation, assess what's needed or possible, and take appropriate steps by relying on (1) their action frameworks (Chapter 6), (2) that framework's inherent ethical dimensions (Chapter 5), and (3) their interactions with, and learning from, their fellow health professionals (both administrators and clinicians) and their patients and families.

It would be incorrect to infer that institutional processes such as the MGB Overarching Goals and Powerful Questions inescapably lead to what we would consider good outcomes. It is common, especially in times of increasingly scarce resources, for hospitals and other health care organizations (including managed care organizations) to become increasingly bureaucratic and cost-conscious. These concerns may lead managers, administrators, and administrative committees to make decisions that clinicians see as running contrary to their own commitments to provide health care to patients.

Even when exactly the same information is available to both administrators and clinicians (as is possible, for example, when administrators and clinicians are working on the same committee), the judgments and preferred outcomes of administrators and clinicians could well differ. As we've mentioned before, there is no reason to expect different clinicians (or, by implication, different administrators) to balance different factors

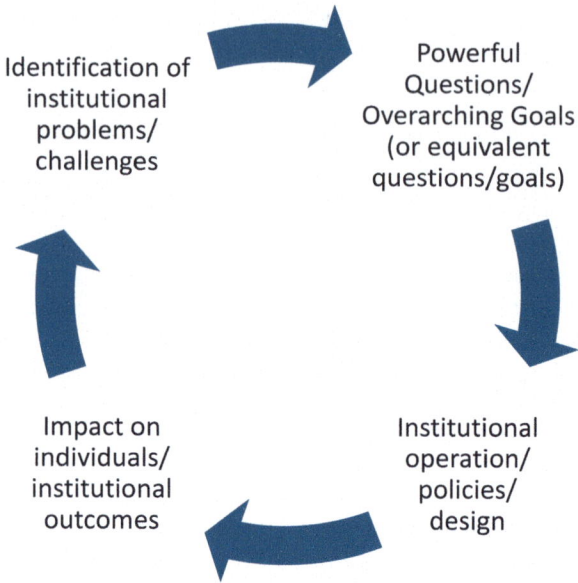

Fig. 11.1 The cycle of institutional change. © Stephen Scher & Kasia Kozlowska 2025

in the same way. But the administrator/clinician divide is different. In reaching a conclusion from, or making a judgment based on, any particular set of facts, administrators would likely balance systemic and financial factors much more heavily than those same factors would be balanced by clinicians, who would presumably come down much harder on the side of optimizing patient care, improving and increasing available resources, and gaining access to the latest technology. In such situations involving group processes, the ultimate outcome will depend upon the capacity of participants to understand divergent perspectives and to reach some sort of productive accommodation.

In some health care organizations, the opportunities for productive accommodation have been lost. As noted, for example, by one experienced primary care physician working in a managed care practice:

Something very deep and very bad has happened, is happening, to medicine. There is so little time, and so little emphasis on spending time with patients, talking to them, asking about their problems, explaining what needs to be done, responding to their fears and wants. It's all a new language: cost efficiency, management talk. This isn't the language of clinical practice that I was trained in. I feel frustrated and very, very alienated. I'm beginning to think it is not for me. I need to get out of it. (Kleinman, 1999, p. 408)

CLINICAL ETHICS AND THE CYCLE OF INSTITUTIONAL CHANGE

There is nothing unique about the process of institutional change described by the MGB Overarching Goals and Powerful Questions. Every institution has its own processes for managing problems and managing change. What is unusual about MGB is the concerted effort it made to articulate those goals and questions—an effort that reflects a strong commitment to ongoing adaptation and progress. Professional schools of health care likewise have their goals and processes, as do training programs within hospitals and other health care settings. That said, what our cycle of institutional change makes clear is that health care institutions—ranging from professional schools to hospitals to myriad other settings—both shape and limit the likely directions of growth except insofar as individuals, teams, or departments (including central administration) step out of their established comfort zones and take strong leadership roles. In such cases, major change, even radical change, is not only possible but potentially welcome.

For decades, health care institutions—including both professional schools and institutions providing care to patients—have undertaken initiatives of one kind or another to improve the ethical behavior of health professionals and the ethical quality of patient care. Our own view is that the ethical model used in these ongoing cycles of institutional change— almost always the "standard approach" described earlier in this chapter—is not up to the challenge presented. Ethics is embedded in every clinical interaction, and any clinically feasible model of clinical ethics needs to be teachable and learnable, clinically relevant, and capable of flexible, adaptable, lifelong use. Such a model must also connect directly and deeply with the needs and perceptions of clinicians. We believe that our socio-institutional approach to clinical ethics does all these things. And in the next chapter we'll try to explain how and why.

NOTES

1. Mass General Brigham is the system that includes Massachusetts General Hospital, Brigham and Women's Hospital, McLean Hospital, and numerous other Harvard Medical School–affiliated institutions. The principles were formulated by a group of clinicians, faculty, and staff working with Mass General Brigham's Office for Strategy and Transformation.

2. These Overarching Goals bear a family resemblance to an institution's or organization's "mission statement," but the latter typically express goals or values that are so broad as to be disconnected from actual decision-making. By contrast, MGB's Overarching Goals are sufficiently concrete to provide a useful framework for institutional policies and related clinical decisions. Note, too, that the Overarching Goals are further specified in the online version of the document (which we do not have permission to reproduce here).

REFERENCES

Accreditation Council for Graduate Medical Education [ACGME]. (2020; implemented July 2021). *Internal medicine milestones*. https://www.acgme.org/globalassets/pdfs/milestones/internalmedicinemilestones.pdf

Dyer, C. (2022). Failure to work collaboratively and learn from incidents led to deaths of babies and mothers at Shrewsbury and Telford trust, review finds. *BMJ (Online), 376*, Article o858. https://doi.org/10.1136/bmj.o858

Fish, S. E. (1980). *Is there a text in this class? The authority of interpretive communities*. Harvard University Press.

Fish, S. E. (1989). *Doing what comes naturally: Change, rhetoric, and the practice of theory in literary and legal studies*. Duke University Press.

Hampshire, S. (1983). *Morality and conflict*. B. Blackwell.

Hojat, M. (2016). *Empathy in health professions education and patient care*. Springer International Publishing.

Kleinman, A. (1995). *Writing at the margin: Discourse between anthropology and medicine*. University of California Press.

Kleinman, A. (1999). Experience and its moral modes: Culture, human conditions, and disorder [lecture delivered in 1998]. In G. B. Peterson (Ed.), *The Tanner lectures on human* values (Vol. 20, pp. 355–420). University of Utah Press; Cambridge University Press. https://tannerlectures.org/lectures/experience-and-its-moral-modes-culture-human-conditions-and-disorder/

Kozlowska, K. (2022). Book review: Psychiatric ethics. *Australasian Psychiatry, 30*, Article 103985622210938. https://doi.org/10.1177/10398562221093874

Mass General Brigham, Office for Strategy and Transformation. (2024a). *Operating principles for how we think and work as an integrated system of care.*

Mass General Brigham, Office for Strategy and Transformation. (2024b). *Some powerful questions for leading our transformation.*

Scher, S., & Kozlowska, K. (2018). *Rethinking health care ethics.* Palgrave Macmillan.

Vize, R. (2022). Ockenden report exposes failures in leadership, teamwork, and listening to patients. *BMJ (Online), 376*, Article o860. https://doi.org/10.1136/bmj.o860

PART IV

The Clinician's Voice

CHAPTER 12

Nurturing the Clinician's Voice

Abstract This chapter highlights the challenges involved in nurturing the clinician's voice, and it also includes some attention to sources of professional dysfunction. We introduce the action cycle as a way of understanding the sequence that begins with one's baseline experience, moves through the touchstone process, and then proceeds to judgment, action, and observation, leading back to baseline experience. We explain how this action cycle fits into clinical training as a means of nurturing the clinician's voice. We conclude by discussing the role of the action cycle in advancing evidence-based health care.

Keywords Action cycle · Baseline human experience · Clinical ethics · Clinical training · Clinician's voice · Evidence-based medicine · *Exit, Voice, and Loyalty* · Health care ethics · Institutional culture · Professional dysfunction · Professional development · Professional education · Program design · Reflective practice · Supervision · Teaching · Time and workload · Touchstone process · Touchstone questions · Touchstones for learning

© The Author(s) 2025
S. Scher and K. Kozlowska, *Revitalizing Health Care Ethics*,
https://doi.org/10.1007/978-3-031-78475-0_12

THE NATURAL HISTORY OF THE CLINICIAN'S VOICE

An easy way of summarizing what we've attempted to achieve in this book is to see it as presenting the *natural history of the clinician's voice*. This history begins with the preprofessional, foundational elements of the clinician's voice, including lifelong familiarity with roles, responsibilities, and their organizational context; the ever-developing dimensions of moral experience and the consolidation of a working moral framework; and the action framework, including all the associated capacities and skills. These elements—which continue, lifelong, to frame and shape how clinicians think, feel, and act—are further shaped by the intense social, institutional, and learning environments characteristic of each field of clinical training. Over the years of professional education and clinical training, the progressive clinical engagement enables clinicians to understand what is, in each particular field of health care, the normal and morally expected course of events, as defined, in part, by what is considered good, competent practice within that field of health care. This understanding enables them, in turn, to identify discrepancies (in relation to past or ongoing actions) and indeterminacies (in relation to prospective actions). And these discrepancies and indeterminacies signal the need, in turn, to review and respond to what has happened or is happening. Whether in the moment (that is, on the spot, as with the Fletchers in Chapter 1) or in a more extended process, clinicians need to sort out the factors in play, take into account the relevant evidence (including the peer-reviewed literature), solve any potential problems, and decide on the path forward. These processes typically require active engagement with, and consideration of, the views of patients, families, and colleagues.

Ethics, or at least the ethics of informal ethical discourse (see Chapters 2, 8, 9, and 11), is embedded in every clinical interaction. Early in the course of professional education, the foundational ethical elements discussed in Chapter 5—autonomy, truthfulness, respect for others, doing good/avoiding harm, solidarity, and moral judgment—all take on the specific interpretations characteristic of each particular field of clinical health care. These interpretations are central to clinicians' understanding of their work and their relations with patients, families, and colleagues. But once students have come to understand these central elements of their clinical role, the progressive growth and elaboration of each clinician's working moral framework is the product of efforts to understand and navigate the clinical environment and to achieve one's aims as a trainee and clinician—including, most importantly, the ongoing demand to provide competent, responsive, timely care to patients. What comes to the fore is that matters of informal ethics are largely embedded in the clinical context—that is, interwoven with "technical" matters relating to clinical skills, processes, and interpersonal interactions.

Taken together, these skills, processes, and interactions are all integral elements of each field's clinical environment, understood as goal-directed, patient-oriented, socio-institutional frameworks that Stanley Fish described as interpretive communities. Each clinician's distinctive voice emerges within and in relation to the complexities of learning and working within any particular interpretive community.

Nurturing (or Not) the Clinician's Voice

Each clinician's voice may have, as we argue, its own natural history, but that's not to say that the processes involved occur effortlessly or without the potential for serious malfunctions. Indeed, quite the reverse is true (Baldwin & Daugherty, 2004; Firth-Cozens & Greenhalgh, 1997; Hojat, 2016; Kleinman, 2019; Nolte et al., 2017). Years are spent in courses where one learns the fundamentals of one's own field, and clinical training adds years to that. The courses can go well or poorly, be well taught or not, be thorough or not, and leave students feeling enthused about, and decently well prepared for, their future clinical work or not. Likewise for clinical training. Consider the following vignette:

> In one of my three-month rotations during my residency in internal medicine, I met my supervisor for the first time on the day that he was expected to sign off on my final evaluation. The only feedback he provided was that "no one has any concerns with your work; everyone says you've been great." While this was reassuring and great for the ego, it was not actually helpful. I would have preferred some assessment regarding my strengths or weaknesses or some form of constructive feedback as to what I should focus on moving forward.

The above vignette is in stark contrast to the following:

> I was a bit anxious at the outset of my fellowship in oncology. Day in and day out, the patients would be medically complex and critically unwell.

The technical and emotional challenges would be constant and potentially overwhelming. Would I be able to cope?

As things turned out, the supervisors went out of their way to support us trainees both intellectually and emotionally. At the orientation sessions, they suggested some core background readings that would help us deal intellectually with the complexities and uncertainties involved with our prospective patients. The sessions also highlighted the emotional dynamics of our work, where we would need to find our own comfortable balance between connecting to the patients emotionally while also maintaining enough distance that we would not be devastated should they die (as many of them would).

To my surprise, the consultants would always make time to ensure that I was learning and coping. They would challenge me to explain what I was doing or planning to do, and why, and then teach me about the next step, the next question to ask. Despite their having a deep and intricate knowledge of the medical care of their patients, the consultants would also ask for my opinion as they knew I knew the patients on the ward the best. And I also enjoyed (and learned from) the questions that they asked when they didn't quite understand how I had reached particular conclusions.

Late that year, as I was struggling to cope emotionally with the slowly approaching death of a patient I had come to know over a period of months, one of the consultants called me into his office. I was sure I had done something seriously wrong. But what I encountered was a cup of hot chocolate with a marshmallow on top, the accompaniment to a long discussion about how to approach death from within a specialty where deaths were common. The consultant also sought my opinion on the best way to manage the patient's individualized care as we entered the palliative phase. A few weeks later there was another hot chocolate and marshmallow waiting for me after the patient passed in hospice.

At the end of that year, not only was I more confident in my knowledge and resilience as a junior doctor, but I possessed a road map for becoming the best senior doctor I could be.

What we see in these two vignettes is the stark contrast between the failure to provide even the most minimal supervision, guidance, or support to a trainee and, at the other extreme, the thoughtful, caring, attentive efforts of senior clinicians to provide a trainee with opportunities for joint reflection and to support the trainee through a difficult, intellectually and emotionally challenging year of advanced clinical training. What is especially noteworthy is the senior clinicians' attention to, and respect for, the trainee as a *whole person*. And it is this attention to the trainee's whole person, including his powerful feelings and ongoing stresses, that both enables him to get through the year and sets him on the path toward long-term growth. What is also noteworthy is the senior clinicians' obvious interest in, and respect for, the trainee's clinical thinking. We suspect that every student and every trainee has had experience at both ends of the spectrum, though the hope is that the productive experiences far outnumber the others.

What strikes us, too, about the second vignette is its integration of the human/ethical and technical dimensions of caring for oncology patients: "how to approach death from within a specialty where deaths were common" and how best "to manage the patient's individualized care as we entered the palliative stage." There's no attempt—nor should there be—to separate out the ethical from the technical dimensions of the questions. They're interconnected, and they're best approached and understood as interconnected. The informal ethical discourse of caring for these patients, or any patients, is embedded in the language of the clinic. It's this language of the clinic that underpins and facilitates how clinicians think, feel, and act. And it's the language that needs to be encouraged and supported from the very outset of professional education in health care.

As the above vignettes suggest, the clinician's voice does not grow and mature on its own. It needs to be nurtured. In our view, the circumstances that work in support of the clinician's voice, as well as those that work to constrain and compromise it, work from the ground up. The development (or not) of the clinician's voice is not a matter of formal didactics but of the lived, concrete, human experience of each student, trainee, and clinician.

SOURCES OF PROFESSIONAL DYSFUNCTION

Before turning specifically to the challenge of nurturing the clinician's moral voice, a short detour will help to put our discussion in context. There are wide-ranging problems within health care today that compromise clinicians' clinical performance, professional competence, and the quality of care that they provide to patients. See Text Box 12.1 for a representative sample of such problems.

Text Box 12.1 Sources of Professional Dysfunction

Inherently Personal Problems

- Personality/personal choices (e.g., laziness, greed, power)
- Psychopathology
- Personal problems (e.g., marital conflict, divorce)

Work-Related Problems

- Chronic lack of support/encouragement/supervision
- Lack of skills needed to cope/adjust with the (potentially changing) circumstances of work
- Dislike of, or impatience with, patients, families, or subsets thereof
- Lack of alignment between personal values/goals and that of the health care system, either broadly or locally
- Blind spots (e.g., in relation to particular illnesses, such as functional neurological disorder [Kozlowska et al., 2021], or patient groups, such as Blacks, Hispanics, women, or indigenous peoples)
- Ongoing power conflicts involving supervisors, administrators, peers, or even patients or families
- Diminishing capacity to cope with recurrent presentations or problems

Consequences of Systemic Problems

- Burnout/compassion fatigue associated with harsh, impersonal, or too-demanding work environment
- Exhaustion, low morale, and difficulties in maintaining a work-life balance in the face of chronic staff shortages
- Chronic stress/distress associated with conflicting cultures of clinicians versus institution (e.g., clinicians' dedication to quality care vs. administration's emphasis on economic efficiencies, time management, and reducing costs, services, and staff)
- Inadequate voice regarding policies and programs (e.g., in the context of an unresponsive administration and potential fear of retaliation)

Some of these problems have their origins outside the health care setting, whereas others are interconnected with the setting and a clinician's capacity to work, or to continue working, within it. And each problem needs to be addressed in its own way. But rather than trying to suggest how any of these problems might be addressed, what we can do is to look at such problems from the perspective of the approach to clinical ethics presented in this volume.

In this context, it's helpful to look at what A. O. Hirschman says about voice in his classic study, *Exit, Voice, and Loyalty* (1970). Hirschman's central idea is that the exercise of *voice*—an expression of which is often taken incorrectly as an ill-willed criticism of the status quo—is actually a form of *loyalty* since the aim is to improve a product, process, or practice within an organization. By contrast, *exit* is a form of *rejection*. For example, the consumer can buy another equivalent product, and the worker can find work elsewhere. But as we noted in our 2018 book, *Rethinking Health Care Ethics* (Scher & Kozlowska, 2018), if trainees or clinicians cannot, in such situations, reasonably exit (which is typically the case), they are in an unwinnable situation unless they can exercise voice without fear of serious consequences.

This bind on trainees and clinicians is centrally relevant to matters of professional dysfunction. If these problems are not explicitly identified, they will simply fester. Formal reviews of one kind or another can only do so much, and even then, such reviews can draw attention to problems only after—and because—they have already had measurable or otherwise observable deleterious effects of one kind or another (Dyer, 2022; Vize, 2022).

It is against this background that the capacity to exercise voice without risk—the organization's institutional culture—comes into prominence. Most cases of professional dysfunction are known by at least some coworkers, and they are the ones in the best position to bring early attention to the dysfunction and consequently to start any potential efforts to intervene and ameliorate the situation.

That's not to say that the decision to intervene is straightforward. Voice is not exercised in a vacuum. To whom does one say something? The person himself or herself? A peer? A supervisor or senior colleague? Surprisingly perhaps, the touchstones for learning and the touchstone questions come into play here. The problem is typically first identified just because something isn't quite right, and then the challenge is to maneuver through the situation via the touchstone questions.

Just who should intervene, plus how and when, often requires careful thought and, indeed, courage. But it must, somehow, be done. In general, doing nothing leaves things to get worse, progressively compromises patient care, and likely influences broader staff morale, too (which often has its own impact on patient care). We should also note that there's a potential inflection point involved here. Detecting a discrepancy or problem and then just letting it pass is a small step toward moral insensitivity and toward distancing oneself from the human dimensions of clinical experience.

THE ACTION CYCLE AND CLINICAL TRAINING

From the perspective of the analysis presented in this book, and separate from the problem of professional dysfunction, there are three crucial stages in the emergence and consolidation of the clinician's voice:

- respect for, and development of, baseline sensitivity to the human dimensions of the clinical setting, as defined by the touchstones for learning: predictive coding, the normal and morally expected course of events, and questions about what we are feeling and why. As noted earlier, these first three steps of the touchstone process are, in the real world, often experienced as one, at least on a conscious level.
- capacity to (a) investigate discrepancies regarding past actions via the touchstone questions in Text Box 7.2 (the last step of the touchstone process) and (b) initiate new actions via the questions in Text Box 9.1.
- capacity to make judgments in responding to discrepancies and in initiating new actions

Figure 12.1 presents these stages as part of an *action cycle* for responding to discrepancies or initiating new actions.

Our view, of course, is that students and especially trainees should be strongly encouraged by their supervisors and other senior clinicians to make use of the touchstones for learning and touchstone questions, and to follow the action cycle as they think about their own experiences and as they encounter the concrete realities of the clinic. Doing so would potentially avoid, for example, the deterioration of human sensitivity to which Arthur Kleinman alludes in *The Soul of Care* (Kleinman, 2019). In particular, he observes that

first-year [medical] students are better at the psychosocial, emotional, and human aspects of taking the patients' medical history. That is to say, there is something untoward, even toxic, in much of the training of doctors that

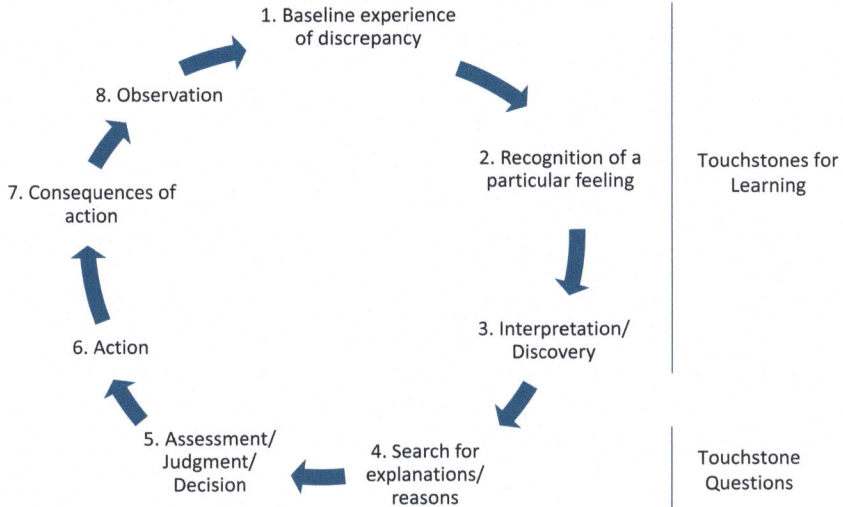

Fig. 12.1 The action cycle in response to discrepancy: Steps (1)–(4) correspond to the *touchstone process*; (1)–(3) correspond to the *touchstones for learning*; and (4) corresponds to the *touchstone questions*. (5) is the subsequent assessment and judgment leading to a specific decision to act. (6) is the action itself, (7) the consequences of that action, and (8) the observation of those consequences, at which point the action in question is judged to be (a) complete, (b) unsuccessful or falling significantly short of expectations, or (c) ongoing but in need of further adjustment. (a) and (b) lead to new, prospective actions and new action cycles (that begin at Step 5 of the action cycle, after addressing the questions regarding prospective action [Text Box 9.1]). The ongoing action in (c) continues, subject to adjustment, leading to the continuation of another cycle for this particular action. © Stephen Scher & Kasia Kozlowska 2025

undermines their social and existential skills at the same time that their technical knowledge and competencies are advancing. (p. 38)

We believe that this point is well taken. It mirrors much of what has been said many times over the years about health care education—and, more broadly, about the "complex demands of an overburdened healthcare system [that places] extraordinary stress on those charged with providing services" (Nolte et al., 2017, p. 4365) (Baldwin & Daugherty, 2004). But we want to move the discussion one step ahead by looking at the problem from the perspective of the *information* lost—the *evidence* lost—if students and trainees do not use the touchstones for learning and touchstone questions, followed by appropriate action, as in the action cycle depicted in Fig. 12.1.

BASELINE SENSITIVITY TO THE HUMAN DIMENSIONS OF THE CLINICAL SETTING

We assume that it would be the rare teacher or clinician who preferred that students or trainees not have access to full information about the care that they are providing to patients. But that's exactly the *inescapable* consequence of students and trainees *not* using the touchstone process—including, most importantly, the touchstones for learning, its first three steps—in perceiving and understanding their clinical encounters with patients, families, and colleagues.

The touchstones for learning are situated, as we've noted, at the intersection of affect and intellect. As such, the touchstones play a pivotal role in maintaining sensitivity to the human dimensions of health care while also leading immediately to a process of inquiry and reflection—the touchstone questions, the last stage of the touchstone process.

But what kind of information is at stake?

Assuming a trainee makes use of the touchstones for learning in providing treatment to, or otherwise caring for, a patient, it is the trainee's predictive coding and understanding of the normal and morally expected course of events that will generate any experience of difference, of discrepancy. It is this baseline experience or information that then needs to be interpreted or investigated.

- If the difference concerns what the trainee sees or detects in a patient, what is it that has changed and that now needs attention? It may be that a specific treatment or intervention has had an observable adverse effect that was not anticipated. It may be that the patient is obviously upset or angry or confused or surprised. Or it may be that the patient is displaying some subtle sign of distress or disquiet.
- If the difference concerns what the trainee sees or detects in a colleague (e.g., a fellow trainee, senior colleague, administrator, or member of the team or other group) who is observing his or her work, what has that colleague observed? Something that the trainee did or didn't do? Some unusual way that the patient reacted or failed

to react? Some subtle sign that something is awry with the patient? Is the trainee responding to some sign, subtle or not, that the colleague is not pleased with what the trainee has done?

– If the difference concerns what the trainee sees or experiences within himself or herself, what exactly is that experience? Frustration? Disappointment? Surprise? Relief? And about what?

One could go on and on with such examples, but the point is as important as it is clear. These human responses of patients and colleagues are sources of information—of *evidence*, even crucially important clinical evidence—about the care and treatment of patients. The same can be said of a clinician's own human responses. As we noted in Chapter 7, the challenge for the trainee is to "interpret" these human responses and to understand their significance in relation to the trainee's expectations and plans—especially since the differences or discrepancies in expectations typically indicate that something has gone wrong, or at least is not as hoped, in the care and treatment of a patient or in communicating with a colleague, patient, or family. Notably, these bits of information—and the further information to which they lead, in turn (see next paragraph)—would likely be disregarded or lost unless trainees attend to their baseline human experience and make use of the touchstones for learning.

The last step in the touchstone process is for trainees to ask the touchstone questions, which seek to understand (with the embedded element of informal ethics) what exactly has happened (or failed to happen), why, and what can be done. And, as with the touchstones for learning, the first three steps of the touchstone process, the touchstone questions generate potentially huge amounts of information that might go unnoticed but for trainees' initial sensitivity to their baseline human experience.

The point here is that the touchstones for learning, especially as integrated into the action cycle presented in Fig. 12.1, should not be characterized simply as "feel-good" questions promoted by clinicians or bioethicists or other scholars or writers who think that modern health care has become too much oriented toward technical skills. These questions at the interface of affect and intellect are important elements in generating and maintaining the evidence base for patient care—in particular, for the patient care that any particular trainee or clinician is providing to patients. To ignore these questions is inescapably to compromise the care and treatment of patients.

EVIDENCE IS EVIDENCE

It has not been our task in this book to present a full-blown analysis of, or approach to, health care education and training. Our goal has been, instead, to provide a natural history of the clinician's voice and to see how that history both precedes and is then interconnected with professional education and training in health care.

Just how that voice develops within the context of professional education and training depends in large part upon the character of that process. The process needs to encourage students and especially trainees to attend to the touchstones for learning—their baseline human experience of clinical health care and their engagement with patients, families, and colleagues—and to then ask and attempt to answer the touchstone questions. When schools and training programs follow such a ground-up process, there is every reason to expect that the voice of clinicians will develop and grow lifelong, much as we saw with the Fletchers in Chapter 1. When trainees' baseline human experience is routinely brushed aside or overridden, learning opportunities are lost, and their capacity for moral sensibility is blunted.

What is also true, however—as discussed above—is that trainees' attention to their baseline human experience actually extends the available evidence and discloses clinical concerns and problems that would otherwise be invisible, with the inescapable consequence that those concerns and problems would remain unaddressed. The care and treatment of patients would be compromised. So, it is an important, though perhaps counterintuitive, consequence of our analysis that students', trainees', and clinicians' attention to, and determination to explore, their baseline human experience in the clinic is *required* as a matter of *evidence-based health care*.

To put this conclusion as concisely as possible: for schools and training programs to integrate this ground-up approach into the design of clinical programs is not, as it were, optional or merely aspirational; it is a *moral imperative* that directly follows from any commitment to pursue evidence-based health care. To ignore this imperative is to ignore fundamental evidence regarding health care delivery and, ultimately, to compromise the care and treatment that patients receive.

To return briefly to our earlier discussion of sources of professional dysfunction within health care, our guess is that, if training programs succeeded in integrating this attention to trainees' baseline human experience into program design, clinicians would actually experience fewer of the problems identified above. They would be much more aware of their own limits and of emerging problems, and much more likely to voice their concerns and to take steps to address problems early and with better outcomes. In this context, clinicians who regularly provide good, human-oriented health care and remain open to their baseline human experience are, in effect, engaged in an excellent form of self-care.

THE PROBLEM OF TIME AND WORKLOAD

We are well aware that making space for the touchstone process and, more broadly, the action cycle (Fig. 12.1) takes real time, and in clinical training, time is a remarkably scarce resource. But *not* making time has real consequences both for trainees and for patients.

Consider the following vignette from Arthur Kleinman's *Soul of Care* (2019) in which he comments on his in-person meeting with a medical resident who had just finished a 24-hour shift with 30 minutes of sleep. The resident

> broke down, crying bitterly as she confronted the stark reality of her situation. Could she live up to her expectations of herself to be a caring physician and still survive the grueling residency? . . . [S]he still could not face the prospect that she would have to put aside what she valued most about medicine, just to get through her training. "If this is what it takes to make it," she said, "will I become a different person when I emerge from training: a doctor who has learned that the only way to resolve this desperate conflict between aspiration and practical reality is to cut corners to get the job done no matter how inadequate the human consequence of patients?" (pp. 213–214)

Time is, indeed, a scarce commodity in training programs and for trainees, but it is also possible for supervisors and senior clinicians to create time for their trainees to reflect about their work. Consider the following progression, with the first coming from a surgeon's memory of his own experience as a surgical fellow (immediately following his residency) and the second coming from the same surgeon years later:

After getting my fellowship I did a huge amount of operating but not once was the consultant present. It was so busy that if I had attempted to "reflect," I would have just fallen asleep.

Years later, as a senior surgeon supervising trainees, this same doctor had come to see

each operation as a learning experience for the surgical trainee, so I made a point of engaging in structured feedback: "What did you think went well?" "What would you do differently next time?" My hope was that reflective analysis such as this, done with the supervisor initially, would eventually be incorporated into the trainee's regular practice and make him or her a better surgeon.

The first remark illustrates the relentless nature of the clinical work coupled with the problem of time and fatigue. And the second points the way toward the pragmatic solution developed by the surgeon: the incorporation of a reflective process into his daily work with trainees. This solution, incidentally, squares with the work of the first author (SS) in teaching clinical ethics in Harvard Medical School–affiliated hospitals), and it also squares with the touchstones for learning, touchstones questions, and action cycle presented earlier in this chapter. The surgeon's first question helps the trainees to identify elements of the surgery that went well and that therefore do not need to be changed (= business as usual/ good clinical care). By contrast, the second presupposes that the trainees have identified something that did not go as well as expected or as hoped (in effect, bringing into play the touchstones for learning, the trainees' baseline assessment of their own experience) and then asks specifically what needs to be changed (which is a touchstone question).

It would be unrealistic to suggest that some senior clinician should always be around to prod trainees into thinking about these baseline human experiences while also trying, moment by moment, to master a technical skill or to complete one task as fast as possible simply in order to get to the next one. But, like the senior surgeon above, any senior clinician can take a moment—for example, during ward rounds or supervision of one kind or another—to query trainees about their baseline assessments of their own work and how to improve it. And training programs can schedule regular time, perhaps an hour each week, in which current clinical problems can be explored with the goal of understanding how such problems are experienced and identified, what might have gone wrong and why, and what needs to be done. In effect, the goal would be to work through, within an hour's time, the full action cycle presented in Fig. 12.1. See Text Box 12.2 for a description of this discussion process.

Text Box 12.2 A Group Discussion Built Around the Action Cycle Depicted in Fig. 12.1

The initial step in the process is to request trainees to select among themselves a case for discussion, with the understanding that the case should be one that is bothering them for reasons unrelated to technical matters of diagnosis or treatment. One resident, nurse, or other trainee or young clinician involved in the case would begin by presenting it to the group (typically 10 to 20 trainees) and framing the initial question or questions for discussion. Insofar as necessary, a senior clinician—including potentially someone with a background in psychiatry, psychology, or even bioethics—should gently help the group sort their way through the potential problems and complexities. The trick here, though, for senior clinicians is to avoid at all costs leading the group along via the route of their own professional expertise. The psychiatrist should avoid psychiatry, and the bioethicist should avoid ethics (especially of the formal kind) except insofar as the trainees frame some part of the discussion in specifically ethical terms. The goal is to find a way forward by respecting, using, and building upon the trainees' own baseline experience, thinking, vocabulary, perceptions, and goals. Insofar as senior clinicians ask guiding questions or make their own observations, it is always helpful to keep the discussion moving along, both back and forth, along the steps of the action cycle.

Both senior clinicians and trainees may object that an hour each week is too much time. Who, in heaven's name, has that much time just to talk? The response is simple: every program director can *make a decision* to schedule an hour a week or every other week for such meetings. And once the meetings are scheduled, every trainee will attend as a matter of course. In addition, the efficiencies and advantages of such a program need to be considered. Maintaining such a program on a regular basis

- expresses a strong institutional commitment to, and interest in, the professional growth and the mental health and well-being of trainees,

- leads to the identification and solution of problems that might otherwise be ignored or, at best, only partially addressed—which may, in turn, ultimately create substantially more work for trainees and other clinicians, and potentially generate avoidable conflict with patients and families,
- enables problems to identified early, before they potentially worsen or go out of control,
- promotes the care and treatment of patients, and
- changes, over time, the intellectual and emotional environment of the program by legitimizing discussion of problems presenting certain sorts of human problems, whether on the part of patients or trainees.

Related to, but separate from, the above points concerning the advantages of regularly scheduled sessions that bring into play the touchstones for learning, much the same advantages can be gained through efforts (including through these same sessions) to improve the listening and communication skills of trainees. A little extra attention early in training can go a long way. A trainee encouraged to use the touchstones for learning will, in listening to patients, pick up discontinuities and discrepancies that other trainees would simply miss. And as is so elegantly communicated in a just-published book by Dean-David Schillinger, *Telltale Hearts* (2024), the stories—personal narratives—that thereby become available to the clinician can change the course of treatment and change patients' lives. Of note, the *quest for narrative coherence*, as discussed in Chapter 6, comes into play here. Unfortunately, this particular capacity is often lost in the demand in clinical training to master skills and to see the patient not as an individual but merely as a bearer of symptoms, ailments, conditions, and diseases. In the process, also lost here is the vast amount of information that better listening and communication skills would generate in efforts to understand patients and families, and to better address their needs.

One final word about evidence. We recognize that instituting clinical conferences as suggested above and endeavoring to improve listening and communication skills have long been understood as a way of promoting both patient care and the mental health and well-being of trainees. That point is a good one. Indeed, most programs recognize that such discussions and other human-oriented interventions are a good thing, but the matter typically stops there. Our point, as noted earlier, is a much stronger one. Any training program committed to evidence-based health care needs to institute such interventions as a matter of both clinical ethics and effective program design.

REFERENCES

Baldwin, D. C., Jr., & Daugherty, S. R. (2004). Sleep deprivation and fatigue in residency training: Results of a national survey of first- and second-year residents. *Sleep, 27*(2), 217–223. https://doi.org/10.1093/sleep/27.2.217

Dyer, C. (2022). Failure to work collaboratively and learn from incidents led to deaths of babies and mothers at Shrewsbury and Telford trust, review finds. *BMJ (Online), 376*, Article o858. https://doi.org/10.1136/bmj.o858

Firth-Cozens, J., & Greenhalgh, J. (1997). Doctors' perceptions of the links between stress and lowered clinical care. *Social Science and Medicine, 44*(7), 1017–1022. https://doi.org/10.1016/S0277-9536(96)00227-4

Hirschman, A. O. (1970). *Exit, voice, and loyalty: Responses to decline in firms, organizations, and states.* Harvard University Press.

Hojat, M. (2016). *Empathy in health professions education and patient care.* Springer International Publishing.

Kleinman, A. (2019). *The soul of care: The moral education of a husband and a doctor.* Viking.

Kozlowska, K., Sawchuk, T., Waugh, J. L., Helgeland, H., Baker, J., Scher, S., & Fobian, A. (2021). Changing the culture of care for children and adolescents with functional neurological disorder. *Epilepsy and Behavior Reports, 16*, Article 1004486. https://doi.org/10.1016/j.ebr.2021.100486

Nolte, A., Downing, C., Temane, A., & Hastings-Tolsma, M. (2017). Compassion fatigue in nurses: A metasynthesis. *Journal of Clinical Nursing, 26*(23), 4364–4378. https://doi.org/10.1111/jocn.13766

Scher, S., & Kozlowska, K. (2018). *Rethinking Health Care Ethics.* Palgrave Macmillan.

Schillinger, D.-D. (2024). *Telltale hearts: A public health doctor, his patients, and the power of the story.* PublicAffairs.

Vize, R. (2022). Ockenden report exposes failures in leadership, teamwork, and listening to patients. *BMJ (Online), 376*, Article o860. https://doi.org/10.1136/bmj.o860

CHAPTER 13

Revitalizing Health Care Ethics

Abstract We begin this concluding chapter by explaining just what we mean by *revitalizing* health care ethics. We see clinical ethics as embodied in clinicians and as maintained through the socio-institutional processes of the interpretive communities that define each particular field of health care. We stress the importance of respecting and building upon the working moral frameworks and action frameworks that all clinicians possess at the outset of their professional education. Finally, we empha-size the role of the touchstones for learning and touchstone questions as a foundation for trainees' and clinicians' lifelong learning and ongoing professional development.

Keywords Action frameworks · Clinical ethics · Clinician's voice · Forms of life · Health care ethics · Informal ethical discourse · Interpretive communities · Lifelong learning · Socio-institutional ethics · Touchstone questions · Touchstones for learning · Wittgenstein · Working moral frameworks

It's a matter of judgment and perspective whether the field of bioethics needs to be revitalized. Some think the field is flourishing; witness the vast amount of resources within hospitals and professional schools of health care now dedicated to bioethics. Others, especially health care clinicians, are skeptical. Notwithstanding potentially useful consultations in difficult

S. Scher and K. Kozlowska, *Revitalizing Health Care Ethics*, https://doi.org/10.1007/978-3-031-78475-0_13

clinical situations, clinicians commonly perceive bioethics as something they learned about as students but that has had little to no impact on how they think about their work or on what they do day in and day out. We acknowledge the difference in perceptions here, but we do not want to be drawn into this particular dispute.

The *revitalizing* that we have in mind takes its meaning from the root of the word itself: the Latin *vita*, for *life*. Our goal is to put ethics back into the lives—literally, into the minds and bodies—of clinicians. Every clinician embodies the values, goals, and standards of their particular field, their particular interpretive community. The community is in them, and the community grows and evolves through them. The community gives life to them as clinicians, and they, as clinicians, give life to the community. Neither can exist without the other.

Socio-Institutional Ethics

It is with this understanding of the process of clinical activity and clinical development that we formulated the notion of socio-institutional ethics—that is, the dynamic, interactive social and institutional processes through which individuals come to be full, functioning members of a professional community. Understanding clinical health care in terms of socio-institutional ethics is not without consequences. What it means is that the ethics of clinical health care, understood as its own phenomenon, develops within a socio-institutional context, and it is this socio-institutional context that gives it both force and ongoing relevance. From the perspective of clinicians, the effect is straightforward: "This is who we are. This is what we do. This is our evidence base. This is how we think. This is how we feel. This is what we believe." Following this line of reasoning, the message to a bioethicist would be the following: "If you want to talk to us, talk to us in our own language and respecting the contours of our world. Your language and your intellectual framework are not necessarily ours."

We are now back to where we started, with Fish's interpretive communities (Fish, 1980, 1989), Hampshire's ways of life (Hampshire, 1983), and Kleinman's moral modes of experience (Kleinman, 1995, 1999). The entire book can be understood as an effort to give shape to these collective ways of thinking about societies, social interactions, professional communities, and, indeed, clinical ethics. With good fortune, we have succeeded decently well and communicated a perspective that explains what we see as the discontinuity between bioethics and, as perceived by clinicians, the informal ethical discourse of clinical health care. But this perspective also points to a way of going forward and creates a window for bioethicists to participate more productively in the world of clinical health care.

Ludwig Wittgenstein, the great Austrian philosopher of the first half of the twentieth century (*Philosophical Investigations* was published posthumously in 1953), wrote of *forms of life* (*Lebensform / Lebensformen*), which were ways of doing things that had no separate justification except within the form of life itself (Wittgenstein, 1953). A form of life is just how things are done. What we are arguing is that the socio-institutional frameworks of clinical health care can be understood as forms of life that need to be understood and criticized in their own terms.

In Part II of this book, we focused on the foundational, preprofessional elements for forging the clinician's voice, including their understanding of roles and responsibilities, along with their working moral frameworks and action frameworks. Though the separate elements of these frameworks are broadly the same for each person, the elements are put together in ways that reflect each person's history, each person's course of development. Moreover, these variations will be largely embedded in, and interconnected with, technical issues and technical language, as well as with matters of individual professional style.

This inescapable variation from clinician to clinician has an important consequence regarding the notion of closure: different clinicians will think about the same clinical situations in different but overlapping ways. Especially in cases of ethical disagreement or conflict, endeavoring to reach a single, ethically correct result—of trying to get every clinician involved to see the situation in the same way and to embrace the same line of reasoning with the same weight given to each factor—is an illusory goal. The *realistic* goal is to find a solution that is "between the flags" (Chapter 8), that all participants in a discussion can accept, with each having their own particular way of understanding the situation and justifying a result. The diversity of viewpoints in such discussions is a strength, and any consensus reached is likely to reflect a workable balance of the factors involved.

LIFELONG LEARNING

In this book we have given extensive attention to the *touchstones for learning* and the *touchstone questions*. One might be tempted to consider this reflective process—the *touchstone process*—as just one way, among others, to start an ethical discussion or to ground a clinician's ethical and clinical thinking. But this process is much more than that: it taps into our brain's capacity to identify something as salient (Barrett, 2017; Pace-Schott et al., 2019), to mark the unexpected via our feelings, including physiological (gut) feelings (Pace-Schott et al., 2019), and to understand particular events as not fitting into the normal and morally expected course of events. These are fundamental skills that trainees and clinicians can use, and that they should be taught and encouraged to use, in every single facet of their work with colleagues, patients, and families. The touchstones for learning, in particular, are a means by which trainees and clinicians can identify and reflect upon their baseline human experience, which serves as a window onto the problems that emerge in the course of their clinical work. Trainees and clinicians will thereby have acquired a skill, a habit, that will stay with them lifelong, and they will also therefore be, lifelong, in a much better position to understand and address the human dimensions of caring for their patients. They will, in effect, be carrying around their own mentors and bioethicists, for they will be asking themselves and their colleagues and students and teams, day in and day out, the same questions that the most effective mentors and bioethicists would be asking them about their work. In the real world, that's a pretty good outcome.

One final word. We hope that clinicians will take to heart the analyses we have presented of working moral frameworks and action frameworks. For each clinician—and uniquely for each clinician—these frameworks for thinking about themselves and others have been developing lifelong and throughout their careers in health care. These are sources of great experience and great strength that should give clinicians confidence and even pride in their capacity, on their own and in consultation with colleagues, to address the ongoing challenges of clinical ethics.

REFERENCES

Barrett, L. F. (2017). The theory of constructed emotion: An active inference account of interoception and categorization. *Social Cognitive and Affective Neuroscience, 12*(11), Article 1833. https://doi.org/10.1093/scan/nsx060

Fish, S. E. (1980). *Is there a text in this class? The authority of interpretive communities.* Harvard University Press.

Fish, S. E. (1989). *Doing what comes naturally: Change, rhetoric, and the practice of theory in literary and legal studies.* Duke University Press.

Hampshire, S. (1983). *Morality and conflict.* B. Blackwell.

Kleinman, A. (1995). *Writing at the margin: Discourse between anthropology and medicine.* University of California Press.

Kleinman, A. (1999). Experience and its moral modes: Culture, human conditions, and disorder [lecture delivered in 1998]. In G. B. Peterson (Ed.), *The Tanner lectures on human* values (Vol. 20, pp. 355–420). University of Utah Press; Cambridge University Press. https://tannerlectures.org/lectures/experience-and-its-moral-modes-culture-human-conditions-and-disorder/

Pace-Schott, E. F., Amole, M. C., Aue, T., Balconi, M., Bylsma, L. M., Critchley, H., Demaree H. A., Friedman, B. H., Gooding, A. E. K., Gosseries, O., Jovanovic, T., Kirby, L. A. J., Kozlowska, K., Laureys, S., Lowe, L., Magee, K., Marin, M. F., Merner, A. R., Robinson, J. L., . . . VanElzakker, M. B. (2019). Physiological feelings. *Neuroscience and Biobehavioral Reviews, 103*, 267–304. https://doi.org/10.1016/j.neubiorev.2019.05.002

Wittgenstein, L. (1953). *Philosophical investigations.* Macmillan.

INDEX

The page numbers followed by 'n', 'f', and 't' represent endnotes, figures, and text boxes.

A

Accreditation Council for Graduate Medical Education
core competencies, 32
Milestones project (internal medicine), 31–40, 33t, 35f, 37f, 38f, 49–50
Action, conceptual distinctions regarding, 80–82, 80t
Action cycle, 192–194, 193f, 197–201, 199t
Action framework (general)
as structuring expectations and responses, 103–107, 120–123
normal and morally expected course of events, 84–87, 93–98, 114–115, 118
predictive coding, 83–84, 93–98, 114–115
salience, 83–84, 114–115
Action framework, individual elements of

capacity to recognize our limits, 88–89
cognitive, perceptual, and affective capacities, 87
communication skills, 88
moral capacities, 87
quest for narrative coherence, 88
theory of mind, 88
Advanced beginner
learning process, 150–157
sample skill (chest drain management), 152–153, 152t

B

Benner, Patricia, 150–152
Between the Flags program, 120–123, 121f, 146, 205
Bioethical expertise, rise of, 48–50

C

Clinician's voice

© The Author(s) 2025
S. Scher and K. Kozlowska, *Revitalizing Health Care Ethics*,
https://doi.org/10.1007/978-3-031-78475-0

foundational elements. *See* Action
    framework (general); Action
    framework, individual elements
    of; Moral experience,
    dimensions of; Roles and
    responsibilities
  natural history of, 186–187,
    192–195
  nurturing, 107–109, 145–147,
    185–201, 193f, 199t
Communities of practice, 27–31

D
Dreyfus, Hubert, 150
Dreyfus model of skill acquisition,
    150–152
Dreyfus, Stuart, 150

E
Ethical principles, 49–50
Evidence-based health care, 192–201
Exit, Voice, and Loyalty, 191–192
Expectations and discrepancies, role in
    professional development,
    149–164

F
Fish, Stanley. *See* Interpretive
    communities
Fletcher, Robert, 3–9
Fletcher, Suzanne, 3–9
Forms of life, 205

G
Generalization of expertise, 48–50
Gilligan, Carol, 62

H
Hampshire, Stuart. *See* Ways of life

Hastings Center/Institute of Society,
    Ethics and the Life Sciences, 48,
    171–172
Hirschman, A.O., 191

I
Individual histories, continuing impact
    and implications of, 46–48, 62
Informal ethical discourse
  compressed language of, 140–143,
    142t
  prospective action and, 131–147
  socio-institutional milieu and,
    125–126
  touchstone questions and,
    113–127, 116t
  vs. formal ethical discourse, 6,
    15–17, 19–25, 167–180
Institutional change, cycle of,
    176–180, 178f
Institutional culture, 176–180,
    189–192
Interpretive communities (Stanley
    Fish)
  defined, 17–18
  socio-institutional processes and,
    172–173

K
Kleinman, Arthur J. *See* Moral modes
    of experience
Kohlberg, Lawrence, 62

L
Lave, Jean. *See* Legitimate peripheral
    participation
Legitimate peripheral participation
    (Jean Lave and Etienne Wenger),
    28–30

Lifelong learning, 160–162, 160t, 206
Lived moral experience, 1–3, 22–23

M
Mass General Brigham (MGB)
  Operating Principles, 175–177
  Some Powerful Questions, 176–177
Merton, Robert K., 17
"Mind the Gap", 50–51
Moral development, theories of, 62
Moral experience, conceptual map of.
  See also Working moral frameworks
  autonomy and related concepts, 66–68
  doing good/avoiding harm, 72–73
  fairness, 63–65
  moral judgment, 75–76
  trust/reliance/respect/reciprocity, 71–72
  truthfulness, 68–70
Moral framework. *See* Working moral frameworks
Moral judgment. *See* Working moral frameworks
Moral modes of experience (Arthur Kleinman)
  defined, 20–22
  socio-institutional processes and, 172–173

N
Normal and morally expected course of events. *See* Action framework (general)
Nozick, Robert, 85

O
O'Brien, Lilian, 80

P
Predictive coding. *See* Action framework (general)
Principles of Biomedical Ethics, 9, 49
Professional dysfunction, 189–192, 190t
Prospective action
  application of policy, 136–137
  changing the course of care, 133–134
  clinical problem-solving, 144–145
  conflict resolution, 137–140
  program design, 134–136
  questions to ask regarding, 132t
  teaching and learning clinical ethics, 145–146
  use of compressed language, 140–143, 142t

R
Rawls, John, 16, 113–114
Reflective practice, 40, 41n, 108, 198–201, 199t
Rethinking Health Care Ethics, 1–9, 23–25, 94–97, 167–168, 191
Revitalizing health care ethics, 2, 203–206
Roles and responsibilities, 52–58

S
Safe and Sound project, 134–136
Salience. *See* Action framework (general)
Segerdahl, Pär, 3
Socio-institutional ethics, 125–126, 172–179, 204–205

T
"Teaching Ethics in Psychiatry: Time to Reset", 3, 94–97, 95t, 101–103

Touchstone process
  defined, 97
  steps, 98f
Touchstone questions
  defined, 96
  examples, 96t, 116t
Touchstones for learning
  defined, 94–97
  examples, 95t

V
Vandenberg, Danielle, 6, 167

W
Ways of life (Stuart Hampshire)
  defined, 19–20
  socio-institutional processes and,
    172–173
Wenger, Etienne. *See* Legitimate
  peripheral participation
Wittgenstein, Ludwig, 205
Working moral frameworks, 75–76,
  87, 103–107, 115–120,
  123–127, 173–175, 186–187

The manufacturer's authorised representative in the EU is Springer
Nature Customer Service Centre GmbH, Europaplatz 3, 69115 Heidelberg,
Germany. If you have any concerns regarding our products, please
contact ProductSafety@springernature.com

Printed and bound by CPI Group (UK) Ltd, Croydon, CR0 4YY
29/04/2026
02099471-0003